D1267823

A NEW VIEW OF CURRENT ACID-BASE THEORIES

A NEW VIEW OF CURRENT ACID-BASE THEORIES

HARMON L. FINSTON
Professor of Chemistry
Brooklyn College, Brooklyn, New York

and

ALLEN C. RYCHTMAN
Analytical Chemist
Glyco, Inc., Williamsport, Pennsylvania

A Wiley-Interscience Publication
JOHN WILEY & SONS
New York Chichester Brisbane Toronto Singapore

Library of Congress Cataloging in Publication Data:

Finston, H. L.
A new view of current acid-base theories.

"A Wiley-Interscience publication."
Includes index.
1. Acid-base equilibrium. I. Rychtman, Allen C. II. Title.
QD477.F56 541.3'94 81-16030
ISBN 0-471-08472-7 AACR2

Printed in the United States of America

10 9 8 7 6 5 4 3 2 1

PREFACE

This monograph is the result of a study based on the premises (1) that all chemical reactions are charge-transfer reactions and (2) that, in aqueous media, the aggregate of water molecules plays a direct role in the transfer of charge. We have critically reviewed the history and development of the accepted acid-base theories and all the varied concepts and classifications with the view that, within their limits, all are valid.

An historical background, which describes the classification of acids and bases from experimental observation, the electrolytic dissociation theory, anhydrobases and aquobases, and pseudoacids, is presented in Chapter 1. Brønsted-Lowry theory is discussed in Chapter 2 along with acidity functions and acidity and basicity in the gas phase in terms of proton affinity. The solvent systems theory is presented in Chapter 3 along with discussions of reactions in protic solvents other than water, aprotic acid-base reactions, relation to electronic structure, acids and bases in melts, and ionotropy. Lewis theory is described in Chapter 4. In Chapter 5 the Usanovich theory is presented, and the relationship that he observed between acid-base reactions and oxidation-reduction reactions is described.

Although the Brønsted-Lowry and Lewis concepts are currently in favor, the most inclusive theory is that of Usanovich, in which oxidation-reduction processes are explicitly regarded as acid-base reactions, and careful examination of the protonic and electronic theories reveals support for a strong connection between acid-base and oxidation-reduction chemistry. However, the Usanovich theory offers no experimental evidence to substantiate this idea, nor does it suggest a general mechanism or specific structure-dependent phenomenon, by which oxidizing and reducing agents manifest their respective acidic and basic tendencies, that is comparable to protolysis in the protonic theory and to coordination in the electronic theory.

The test for the hypothesis that oxidizing agents are acids is described in Chapter 6. Conventional methods of measuring acidity, that is, pH measurements and acid-base titrations, were applied to determine the acidity of oxidizing agents. Also described are paper-strip electrophoresis experiments that suggest a mechanism by which oxidizing agents manifest acidity.

Dichromate ion (or, more correctly, the dichromate-bichromate mixture) behaves like a weak acid in aqueous solution and, apparently, decreases pH only slightly with increasing concentration, but this visible change is only a small fraction of the actual increase in acidity, which must be sufficiently large to overcome the Debye-Hückel effect and the effect of changing liquid-junction potential, both of which would tend to raise the pH. Paper-strip electrophoresis experiments with Cr(VI) species in supporting media of varying pH reveal the existence of what appear to be hitherto unreported species of relatively low ionic mobility predominant at low pH ($pH<2$).

It is proposed that the usual Grotthus mechanism for rapid proton transfer in aqueous solutions is altered by the presence of $Cr_2O_7^{2-}$ and $HCrO_4^-$ and that "hydroniumation" of these species occurs. This phenomenon accounts for both the increased acidity observed in the pH measurements and the low mobility observed in the electrophoresis experiments.

We are grateful for the helpful discussions and suggestions provided by Orest Popovych, Joseph Glickstein, and Evan T. Williams of the Department of Chemistry, City University of New York, Brooklyn College and Robert Kirby of the Department of Chemistry, City University of New York, Queens College. One of the authors (H.L.F.) acknowledges his great debt to Taitiro Fujinaga, Department of Analytical Chemistry, University of Kyoto, for his initial encouragement.

HARMON L. FINSTON

Brooklyn, New York

ALLEN CARL RYCHTMAN

Williamsport, Pennsylvania
December 1981

CONTENTS

CHAPTER

1

HISTORICAL BACKGROUND

For three centuries the nature of acids and bases, and their relationship has been the subject of scientific investigation. It is hoped that this book will contribute to the resolution of the continuing debate over acid–base concepts.

1.1 EARLY ACID–BASE IDEAS

The word "acid" was derived from "acetum," the Latin word for acidic plant juices such as vinegar. "Alkali," the predecessor of "base," came from "al kalja," the Arabic word for plant ashes.

Early acid–base classifications resulted from experimental observations.[1–6] Acids were recognized by their sour taste, solubility in water, and ability to dissolve other substances. Boyle added to this list the abilities to precipitate sulfur from sulfide solutions and to turn blue plant dyes red. Acids lost these distinctive traits upon contact with alkalies. The latter also possessed identifying characteristics, often in direct contrast to those of acids: they dissolved sulfur and oils, felt slippery, turned plant dyes blue, and lost these traits upon contact with acids.

In 1648 Glauber recognized the contrasting properties of acids and alkalies as the consequence of a fundamental "oppositeness" of the two.[7] Glauber was also the first to state that a salt was the product of an acid–alkali reaction. In 1744 Rouelle applied the salt-formation concept to reactions between acids and an enlarged class of substances that he termed "bases"; the new category included alkalies (the hydroxides and carbonates of sodium and potassium), their alkaline-earth equivalents, metals, and some oils.[1,7] The word "base" was derived from the observation that the reaction products of these substances with acids sometimes decomposed upon heating into a volatile acidic component and a nonvolatile basic (in the sense of fundamental) residue,[3] for example,

$$Na_2CO_{3(s)} \rightarrow \underset{\text{Base}}{Na_2O_{(s)}} + \underset{\text{Acid}}{CO_{2(g)}} \quad (1)$$

Two years later William Lewis also concluded that the typical reaction between acids and certain other substances was salt formation.[7]

In the mid-seventeenth century Boyle attempted to attribute the properties of acids, especially the sour taste, to what he believed were sharp-edged acid particles. Lemery expanded Boyle's idea by visualizing bases as porous substances, and acid–base interactions as the penetration of sharp acids into porous bases.

There was a shift in emphasis during the eighteenth century from attempts to interpret properties of acids to the search for an "acidifying principle," and element or combination of elements common to all acids.[2] Most chemists had adopted Rouelle's view that bases were species of varying composition, having little in common with each other except the ability to react with acids to form salts; so no concurrent search for a "basic principle" was under way and research into the nature of bases aroused little interest for the next century.

In 1774 Priestley put forth the view that carbon dioxide ("fixed" or "dephlogisticated" air) was the acidifying principle, basing this idea on the ability of CO_2 to react with "phlogiston" from the combustion of other substances. The discovery that carbon and oxygen were constituents of all organic acids seemed to substantiate Priestley's viewpoint, but the concept found only limited acceptance.

The first widely accepted theory of acids emerged from the oxidation experiments of Lavoisier, who found that common acids were often the combustion products of nonmetals and oxygen. In 1777 he proposed the element oxygen* as the long-sought acidifying principle and a necessary constituent of all acids.[1-7] An acid was defined as a compound of oxygen and a nonmetal, for example, carbon, nitrogen, phosphorus, or sulfur. Salt formation was considered to be the characteristic acid–base reaction.

Lavoisier was sure enough of his conclusions to rename muriatic (hydrochloric) acid "oxymuriatic acid" and to insist that it contained oxygen, without any supporting experimental evidence. Davy attempted to provide such evidence in numerous experiments performed from 1810 to 1815 but, unable to detect any oxygen, became convinced that muriatic acid contained no oxygen and that oxygen was not the acidifying principle. This conclusion

*The word "oxygen" ("acid former") was invented by Lavoisier.[2,3]

was confirmed by others for a variety of acids over the next few years, and by 1830 the existence of oxygen-free acids was well established.

Despite its disproval, the impact of Lavoisier's early attempt to systematize the chemistry of acids can neither be minimized nor denied. In a sense the oxygen theory remains with us in the fact that the words for "acid" and "oxygen" have the same root in such diverse languages as German ("Säure" and "Säuerstoff"), Russian ("kislota" and "kislorod"), and Hebrew ("chumtzah" and "chamtzan").[6]

Disproof of the oxygen theory again left chemists without an acidifying principle, and during the two decades following Davy's experiments the nature of acids became a topic of sharp controversy.[1–7] Davy reasoned in 1815 that no single acidifying principle existed and that acidity was a consequence of the arrangement of elements in substances rather than the presence of a particular element. However, he abandoned this viewpoint the following year, rejecting anhydrous iodine pentoxide as an acid. The elemental composition of water was well established by this time, and Davy's requirement that iodine pentoxide combine with water before it could behave as an acid led him to believe that hydrogen was the necessary constituent of all acids. Dulong reached the same conclusion in 1815 from his determination of hydrogen in oxalic acid and from his observation that hydrogen-containing acids reacted with metallic (basic) oxides to form anhydrous salts and water.

Many did not accept the ideas of Davy and Dulong, because to do so was to exclude substances with pronounced acid properties, for example, anhydrous sulfur trioxide, from classification as acids, although their hydrates, for example, sulfuric acid, were considered acids. The modified oxygen theory of Gay-Lussac attributed the acidic properties of oxyacids to oxygen, those of hydrogen iodide to iodine, those of hydrogen cyanide to cyanide, and so on. Gay-Lussac labeled the oxygen-free acids "hydracides." In 1826 Bonsdorff provided proof of Gay-Lussac's contention that behavioral similarities existed between nonmetallic oxides and halides by demonstrating the acidic reaction of litmus with the latter. Bonsdorff also found that alkali chlorides turned litmus blue, and he attributed limited basic properties to them. Further experiments led him to propose the existence of "acid-analogous" and "base-analogous" classes of substances that reacted with litmus in a manner identical to those of acids and bases. Mercuric and platinic chlorides were acid analogues, as was boron trifluoride, while base analogues included the alkali halides.

Berzelius also believed in a modified oxygen theory, but combined it with his concept of dualistic electrochemistry, a forerunner of the ionic theory that became prominent a half-century later. Berzelius' objection to Gay-Lussac's ideas was based on the absence of a satisfactory explanation for the strong acid properties of HCl compared to water, a neutral substance despite the presence of the more electronegative oxygen atom. Berzelius defined an acid as a substance with excess negative electricity, and a base as one with electropositive character, dispensing with Bonsdorff's acid- and base-analogous categories considering the substances included therein as acids and bases in their own right. Thus the Berzelius definition of an acid included the halides of boron, mercury, hydrogen, and nonmetals as well as nonmetallic oxides; metallic oxides and halides were bases. Acid–base reactions were envisioned as the neutralization of electropositive (basic) particles by electronegative (acidic) particles,

$$\underset{\text{Base}}{Na_2O} + \underset{\text{Acid}}{2HCl} \rightarrow \underset{\text{Neutral salt}}{2NaCl + H_2O} \tag{2}$$

or, in terms of the atoms actually carrying the electropositive and electronegative character,

$$Na + Cl \rightarrow NaCl \tag{3}$$

Another process included in Berzelius' concept of acid–base reactions was the combination of boron halides with alkali halides:

$$\underset{\text{Base}}{KF} + \underset{\text{Acid}}{BF_3} \rightarrow \underset{\text{Neutral salt}}{KBF_4} \tag{4}$$

The proposals of Berzelius failed to stem the increasing scientific acceptance of the Davy–Dulong hydrogen viewpoint of the 1830s. In 1838 Liebig defined an acid as a hydrogen compound in which the hydrogen could be replaced by a metal, and by 1840 this view was almost universally accepted, remaining so for over 40 years. No corresponding constitutive base theory evolved during this period; bases were regarded merely as antiacids, that is, substances capable of reaction with acids to form salts.[6,8] In addition, none of the aforementioned acid or acid–base concepts presented anything more than a qualitative picture of acid–base behavior.

1.2 ELECTROLYTIC DISSOCIATION THEORY

The theory of electrolytic dissociation, which attributed reactions in solution to the presence of ions, was largely developed by Arrhenius and Ostwald between 1880 and 1890. Arrhenius applied the theory to acids and bases in 1887, defining an acid as a hydrogen-containing substance that dissociated into hydrogen ions and anions when dissolved in water. A base was defined as a hydroxyl-containing substance that dissociated into hydroxide ions and cations in aqueous solution. Neutralization, the reaction of an acid with a base, became the combination of hydrogen and hydroxide ions to form water, accompanied by salt formation.

The examples of equations (5)–(7) summarize the Arrhenius definitions:[1,3-8,10]

$$\text{Acid:} \qquad HCl_{(aq)} \rightarrow H^+{}_{(aq)} + Cl^-{}_{(aq)} \tag{5}$$

$$\text{Base:} \qquad NaOH_{(aq)} \rightarrow Na^+{}_{(aq)} + OH^-{}_{(aq)} \tag{6}$$

$$\text{Neutralization:} \qquad H^+{}_{(aq)} + OH^-{}_{(aq)} \rightleftharpoons H_2O \tag{7}$$

The Arrhenius acid–base theory was an improvement over previous ideas into two respect: bases were defined on a constitutive premise and not merely as substances reacting with acids; and the Arrhenius concept also presented the first quantitative picture of acid–base behavior. An aqueous solution containing more hydrogen than hydroxide ion was acidic; a solution in which the reverse was true was basic; and one with equal concentrations of both ions was neutral.

The electrical conductivity of a solution, a consequence of ionic dissociation, provided a convenient means of measuring the strengths of acids and bases. Since conductivity increased with increasing dissociation, acid (or base) strengths could be quantitatively compared by measuring the conductivities of their aqueous solutions at equal analytical concentrations. Completely dissociated acids and bases, for example, HCl and NaOH [equations (5) and (6)], exhibited high conductivities in aqueous solution and were therefore regarded as strong acids and bases. The low conductivities of solutions of incompletely dissociated acids and bases were attributed to their weakness. Acid–base dissociation was depicted as a dynamic equilibrium process in which the rate of dissociation of electrically neutral molecules equalled the recombination rate of their ions:

$$HA \rightleftharpoons H^+ + A^- \tag{8}$$

$$MOH \rightleftharpoons M^+ + OH^- \tag{9}$$

Application of the mass-action law to equations (8) and (9) yielded ionization or dissociation constants (K_a for acids, K_b for bases):

$$K_a = \frac{[H^+][A^-]}{[HA]} \tag{10}$$

$$K_b = \frac{[M^+][OH^-]}{[MOH]} \tag{11}$$

where the brackets represent the concentrations of the bracketed species. Equations (10) and (11) are related to the degree of ionization α of the acid or base and its original analytical concentration C:

$$K_a = \frac{\alpha_{\mathrm{HA}}^2 C_{\mathrm{HA}}}{1-\alpha_{\mathrm{HA}}} \tag{12}$$

$$K_b = \frac{\alpha_{\mathrm{MOH}}^2 C_{\mathrm{MOH}}}{1-\alpha_{\mathrm{MOH}}} \tag{13}$$

The acid and base dissociation constants derived by relating conductivity to the degree of ionization furnished a means of quantifying the acid or base strength in aqueous solution, useful in acidity measurements, titrations, the preparation of buffers (solutions of weak acids or bases and their salts), and they correlated well with estimations of relative acid and base strengths from indicator and catalytic methods.

The Arrhenius acid–base concept was also the first to account for the fact that salts produced as a result of acid–base neutralization were often not neutral but, rather, exhibited acidic or basic properties in water. For example, NH_4Cl, the product of HCl and NH_4OH (utilizing the strict Arrhenius formula for the base), manifested weak acid properties in aqueous solution, whereas sodium acetate, produced by the reaction of NaOH and acetic acid, behaved like a weak base. Traditionally these phenomena had been explained by the predominance of the properties of one parent in the acid–base pair from which the salt was formed; for example, in NH_4Cl the strong acid properties of HCl predominated over the weak basic properties of NH_4OH, and in CH_3COONa the properties of the basic parent were dominant. The Arrhenius concept visualized these processes as hydrolysis reactions, that is, reactions of acidic or basic salts with water to regenerate parent acids and bases:

$$NH_4Cl_{(aq)} + H_2O \rightarrow NH_4OH_{(aq)} + HCl_{(aq)} \tag{14}$$

$$CH_3COONa_{(aq)} + H_2O \rightarrow CH_3COOH_{(aq)} + NaOH_{(aq)} \tag{15}$$

These reactions were the reverse of the neutralization process and, like neutralization, also depended on the dissociation constants of water and of the acid–base pair involved. The properties of the stronger (more ionized) parent predominated.

Although superior to its predecessors, the Arrhenius acid–base theory had many shortcomings:

1. Acids and bases were not defined as pure substances but, rather, in terms of aqueous solutions of pure substances. For example, pure anhydrous hydrogen chloride was not an acid and manifested no acidic properties according to Arrhenius, whereas aqueous HCl was regarded as an acid. Similarly, pure ammonia was not considered a base, but aqueous NH_3 ("ammonium hydroxide") was regarded as a base.

2. Acid–base behavior was not acknowledged in nonaqueous solutions, despite observations of such behavior in solvents such as liquid ammonia.[1,7–9] Reactions occurring in the gas phase or fused state were likewise not considered acid–base phenomena.[1,4]

3. The theory also required that ionization be a necessary characteristic of acids and bases, a consequence of the general electrolytic dissociation theory of reactions in solution. This caused confusion when the question of acidity in nonionizing solvents arose, for example, although hydrogen chloride ionized completely in water, it remained practically undissociated in benzene, toluene, anhydrous sulfuric acid, and in the pure liquid state, making it unclear whether HCl was considered an acid in circumstances other than aqueous solution according to the Arrhenius theory, on the basis of both the aqueous solution and ionization limitations.

These points emphasize the limitation of the Arrhenius theory of acids and bases to merely a definition of an imbalance of the ions of one particular solvent, contrary not only to intuition but also to experimental observation. Acid–base reactions[12] and catalytic activity[28] were evident in solvents in which very little, if any, ionization occurred. A small minority of chemists had never accepted the electrolytic dissociation theory as universally applicable to inorganic reactions and amassed evidence by demonstrating that reactions between inorganic species in nonionizing solvents were often rapid.[11] The low conductivity observed throughout such processes implied that ionization was not a prerequisite for all inorganic reactions. Folin and Flanders[12] in 1912 extended this idea directly to acid–base reactions by titrating HCl, H_2S, and several organic acids with sodium ethoxide or sodium amylate to a phenolpthalein end point in benzene or chloroform. Their experiments also involved solutions of low conductivity, that is, low degree of ionization, yet neutralization proceeded rapidly and end points were sharp, convincing Folin and Flanders that ionization was not necessary for acid–base reactions. However, the electrolytic dissociation theory was then widely accepted and the small body of evidence in contradiction was largely ignored.

4. Acids remained a class restricted to hydrogen-containing compounds, as had been the case since the time of Liebig. Substances with obvious and sometimes pronounced acidic properties that did not contain hydrogen, for example, SO_3, $SnCl_4$, and CO_2, continued to be excluded.

5. Although Arrhenius' definition of bases represented an advance over previous ideas, many substances containing no dissociable hydroxyl group were recognized as having a basic character and the ability to neutralize acids. Organic chemists, many of whom had never accepted Arrhenius' definition, continued to demonstrate the basicity of ammonia, amines, aniline,[13] alcohols,[4,18] pyridine and its derivatives,[14] quinones,[15] and so on. A strict interpretation of the Arrhenius concept excluded even metallic oxides from being classified as bases.[4] Advocates of the Arrhenius theory rationalized the basicity of metallic oxides, ammonia, and amines by including hydration in their concept, although this raised the question of whether basicity was a property of the substance itself or of its hydrate[8]:

$$K_2O + H_2O \rightarrow 2KOH \tag{16}$$

$$CaO + H_2O \rightarrow Ca(OH)_2 \tag{17}$$

$$NH_3 + H_2O \rightarrow NH_4OH \tag{18}$$

$$NR_3 + H_2O \rightarrow NR_3HOH \tag{19}$$

6. An increasing quantity of evidence disputed the existence of free hydrogen ion in water and other solvents. Goldschmidt[16,17] observed a 25-fold decrease in the rate of HCl-catalyzed esterification of organic acids in anhydrous alcohol upon addition of a few percent of water and attributed this effect to the greater degree of HCl dissociation in water compared to ethanol, determined by conductivity measurements. Theorizing that the free hydrogen ion of HCl present in anhydrous alcohol became bound to added water and existed as the hydronium (or oxonium or hydroxonium) ion, H_3O^+, Goldschmidt accounted for his data by assuming that the hydronium ion was the carrier of catalytic and other acid properties in water and that the catalytic power of the hydronium ion was weak compared to that of free hydrogen ion.

Fitzgerald and Lapworth[18] agreed that very little, if any, free hydrogen ion existed in aqueous solution, but did not believe that H_3O^+ possessed any catalytic power, interpreting Goldschmidt's observations to be caused not by replacement of the strong catalyst H^+ by the weak H_3O^+ but, rather, by a simple decrease in the concentration of the only true catalyzing acid, the free proton. Formation of a dissociating hydronium salt superseded simple dissociation as the distinctive reaction of acids in water [cf. equations (5) and (8)]:

$$\underset{\text{Acid}}{HCl} + H_2O \rightarrow \underset{\text{Hydronium salt}}{H_3O^+Cl^-} \rightarrow \underset{\text{Dissociated hydronium salt}}{H_3O^+ + Cl^-} \quad (20)$$

$$CH_3COOH + H_2O \rightleftharpoons CH_3COO^-H_3O^+ \rightleftharpoons CH_3COO^- + H_3O^+ \quad (21)$$

Fitzgerald and Lapworth therefore regarded conductivity measurements in aqueous solutions to be of limited value in determining acidity, since the species being determined in such measurements was the cation of a salt (H_3O^+), not that of an acid (H^+).

Lapworth later mathematically demonstrated that, in principle, acid properties could be explained without resorting to the free proton concept.[19] The properties of an acid dissolved in a basic solution depended, in general, upon the extent of salt formation (acid–base combination), the extent of acid partition between competing bases, and the dissociation of the resulting salts. For the specific case of an acid dissolved in water, acidity was a function of the acid's affinity for water and of the dissociation of the resultant hydronium salt. Lapworth eventually concluded that free protons did not exist, either in alcohol or in water, and suggested that undissociated acids played a role in acid catalysis, implying the that free protons did not exist in any solvent.[20]

These ideas further emphasized the weaknesses of the Arrhenius theory. Goldschmidt's proposal[16,17] that free protons did not exist in water was adapted to other ionizing solvents,[20] with dissociation reactions formulated analogously to equations (20) and (21), which are employed to describe acid–solvent interactions. The hydronium salt formation concept clearly illustrated the basic properties of water, in sharp contrast to the Arrhenius depiction of water as a perfectly neutral solvent and in close agreement with the observations of Hantzsch (see Section 1.4). The equation for acid hydrolysis was modified [cf. equation (14)]:

$$NH_4Cl + 2H_2O \rightarrow H_3O^+ + Cl^- + NH_4OH \quad (22)$$

Although the shortcomings of the Arrhenius theory were well known almost from the time of its inception, the theory managed to retain a wide following until the advent of the Brønsted–Lowry theory in the 1920s, mainly because of the general prominence of ionic theories and relative paucity of nonaqueous acid–base studies during the latter part of the nineteenth and the early part of the twentieth century. Vorlanders's[21] unsuccessful attempt at expanding the Arrhenius definitions to nonaqueous solutions restated them in terms of Liebig's definition: an acid was a hydrogen-containing compound in which all or some of the hydrogen was replaceable by a metal, and a base was an oxide or a hydroxyl compound in which all or some of the oxygen or OH was replaceable by an acid radical.

1.3 ANHYDROBASES AND AQUOBASES

The classical Arrhenius concept pictured an acid–base reaction as a metathesis producing water and an ionic salt:

$$HX + MOH \rightleftharpoons M^+X^- + H_2O \tag{23}$$

According to Arrhenius, the extent of such a reaction depended on the ionization of the reacting acid–base pair and of water [equations (7)–(9)].

Pfeiffer[22,23] and Werner[24] discovered that equation (23) did not hold true for hydroxyl-containing transition metal compounds in which the hydroxyl group was not ionizable, although these compounds reacted with acids. Moreover, the products of such reactions were often less ionic than expected, and sometimes not ionic at all. In his coordination theory Werner distinguished between two types of bonding in complex compounds: nonionic bonding of ligands to central metal atoms or ions in a primary coordination sphere and ionic bonding of species outside the coordination sphere. Werner also noted the general importance of hydration in increasing the ionic character of complex compounds, for example,

$$[CoCl(NH_3)_5]Cl_2 + H_2O \rightleftharpoons [Co(H_2O)(NH_3)_5]Cl_3 \tag{24}$$

and its specific role in transforming coordinately bound hydroxocomplexes into ionically bound aquocomplexes, for example,

$$\underset{\text{Hydroxocomplex (nondissociating OH)}}{[Co(OH)(NH_3)_5]Cl_2} + H_2O \rightleftharpoons \underset{\text{Aquocomplex (dissociating OH)}}{[Co(H_2O)(NH_3)_5](OH)Cl_2} \tag{25}$$

Pfeiffer and Werner both used the hydration model to explain acid–base reactions involving transition metal hydroxocomplexes. Instead of occurring via ionization followed by metathesis [equation (23)], these processes involved incorporation of the entire acid molecule into the base in a manner similar to hydration. The products of these addition reactions were oxonium-type aquometal salts,

$$M'OH + HX \rightleftharpoons M'OH_2{}^+X^- \tag{26}$$

that subsequently eliminated water, leaving as a final product a compound of low or no ionic character,

$$M'OH_2{}^+X^- \rightleftharpoons M'X + H_2O \tag{27}$$

A specific example of equations (26) and (27) further elucidates the concept:

$$[Co(OH)_3(NH_3)_3] + 3HCl \rightleftharpoons [Co(H_2O)_3(NH_3)_3]Cl_3 \tag{28}$$

$$[Co(H_2O)_3(NH_3)_3]Cl_3 \rightleftharpoons [CoCl_3(NH_3)_3] + 3H_2O \tag{29}$$

Pfeiffer drew an analogy between reactions such as equations (26)–(29) and those of other basic species, for example, amines:

$$RNH_2 + HX \rightleftharpoons RNH_3{}^+X^- \tag{30}$$

$$RNH_3{}^+X^- \rightleftharpoons RX + NH_3 \tag{31}$$

The hydroxocomplex acid–base [equations (26) and (28)] and hydration [equation (25)] processes were believed to occur via a proton-transfer mechanism involving direct proton transfer from an undissociated water molecule rather than proton addition following the ionization of water, as advocates of the Arrhenius concept believed. Werner[24] reached this conclusion by considering the improbability of the dependence of rapid hydration reactions on the small ionization constant of water. Direct substitution of an undissociated water molecule for a hydroxyl ligand was rejected as unlikely in view of the acid–base reactions [equations (26) and (28)].

The postulation of a proton-transfer mechanism led both Pfeiffer[23] and Werner[24] to disregard the ability to produce hydroxide ion as the definitive acid–base property, since complex hydroxobases often contained no ionic hydroxide, although some possessed very strong basic properties, as Werner had determined qualitatively from tests with litmus and a variety of acidic species, for example, $AgNO_3$, CO_2, $NH_4{}^+$, CH_3COOH, and mineral acids. The affinity of a base for protons (of water or any other protogenic substance) was recognized as the decisive factor in determining base strength:

$$M'OH + H^+ \rightleftharpoons M'OH_2{}^+ \tag{32}$$

Hydrolysis of metal salts was no longer regarded as a reaction between two independent species, a salt and a water molecule [equation (23)], but, rather, as the dissociation of an aquometal ion [equations (26) and (32)]. The ionization of water, so important in Arrhenius theory, was relegated to a position of no significance.

Pfeiffer[22] distinguished between "pure" bases, reacting in accordance with the classical Arrhenius concept [equation (23)], and transition metal bases, which he named "pseudobases" because their reaction with acids led to the formation of nonionic "pseudosalts." Werner[24] made no such distinction; generalizing the concept, he perceived all bases as capable of existence in an "anhydro" and an "aquo" form. Anhydrobases contained no dissociating hydroxide and reacted with water to form aquobases. The latter were active basic species, capable of dissociating into hydroxide ion and a complex cation [cf. equation (24)]:

$$\underset{\text{Anhydrobase}}{KOH} + H_2O \rightleftharpoons \underset{\text{Aquobase}}{[KOH_2]OH} \tag{33}$$

$$NH_3 + H_2O \rightleftharpoons [NH_4]OH \tag{34}$$

Werner conceived of all acid–base reactions as producing aquosalts, some of which were unstable and eliminated water, ending up as anhydrosalts [cf. equations (26)–(29)],

$$[KOH_2]OH + HCl \rightarrow [KOH_2]Cl + H_2O \tag{35}$$

$$[KOH_2]Cl \rightarrow KCl + H_2O \tag{36}$$

thus accounting for the fact that many simple salts contained no water of hydration. Werner eventually extended, by analogy, his anhydro–aquo concept to acids:

$$HX + H_2O \rightleftharpoons H[XH_2O] \tag{37}$$

The hydration of simple acids and bases may be viewed as an unnecessary encumbrance that did not clarify simple acid–base phenomena and imposed the aqueous solvent restriction on acid–base reactions. However, the establishment of proton affinity, rather than the ability to produce hydroxide ion, as the criterion for determining basicity represented an important development that was to serve as a partial basis for the Brønsted–Lowry theory and is still considered significant, in spite of the appearance of more general acid–base concepts.

1.4 PSEUDOACIDS

In 1899 Hantzsch[6,8,25] noticed that certain organic nitrocompounds manifested acid properties and were neutralized by strong bases in aqueous solution but, unlike most neutralization processes, these reactions proceeded at a measurably slow rate. This was evident from the gradual increase in conductivity with time of the acid–base mixture; eventually the conductance reached a constant value, signifying completion of the reaction. A slow rate of reaction was also observed for the reverse process, that is, the addition of a strong acid to a solution of a nitro-organic salt to regenerate the original nitro-acid. A gradual decrease in solution conductivity accompanied this reaction.

These findings were not related to the low acid strength of the nitro-organics, because other weak acids, for example, acetic acid, reacted instantaneously with bases. Hantzsch concluded that the nitro-organic acids existed in two isomeric forms, an electrolyte and a nonelectrolyte one, related to each other by the intramolecular shift of a labile hydrogen atom. Conductivity measurements indicated that most nitro-organics existed as nonelectrolytes. Hantzsch therefore proposed that the hydrogen shift converting the nonelectrolyte into a dissociating acid occurred prior to reaction with a base and that the shift was a slow process, accounting for the observed rate of reaction. For nitromethane, the simplest nitro-organic, Hantzsch's proposals may be represented by

$$\underset{\text{Nonelectrolyte (pseudoacid)}}{CH_3NO_2} \rightleftharpoons \underset{\text{Electrolyte (true acid)}}{CH_2NOOH} \qquad \text{(slow)} \tag{38}$$

$$CH_3NOOH + OH^- \rightarrow CH_2NOO^- + H_2O \qquad \text{(rapid)} \tag{39}$$

Hantzsch reversed equation (38) to explain the slow rate of reaction of a nitro-organic salt with a strong acid on the premise that hydrogen shifts were slow in both directions. He referred to the nonelectrolyte isomers as "pseudoacids" and employed the term "true acids" to describe the reactive electrolyte isomers. Although his original experiments involved measurably slow reactions, Hantzsch realized that the relative nature of slowness precluded reaction rate as a criterion for distinguishing between true and pseudoacids. An intramolecular shift was the decisive factor and the original pseudoacid classification[25] included not only acidic species undergoing slow neutralization but any acidic species showing evidence of an intramolecular shift, for example, keto–enol tautomers, weak acids whose salts hydrolyzed slowly or not at all, colorless acids whose salts or anions were colored, and so on.

Hantzsch expanded his definitions into a general theory on the nature of acidity over the next three decades. Extensive investigations of acid behavior in the pure state, in water, and in nonaqueous solvents led him to classify acids as true or pseudo under varying sets of experimental conditions according to the following definitions[26,27]:

1. An acid was a hydrogen-containing compound in which hydrogen was replaceable by a metal. This apparent regression to Liebig's concept defined acids in terms of salt-forming ability, which Hantzsch considered the definitive acid property, rather than in terms of the existence of free or solvated hydrogen ions.
2. A true acid was a heteropolar hydrogen compound containing ionically bound, active hydrogen and was capable of undergoing direct salt formation without any structural or appreciable spectral change. The strength of a true acid lay in its salt-forming power when undissociated.
3. A pseudoacid was a homopolar hydrogen compound containing nonionically bound, inactive hydrogen and was able to undergo salt formation only with structural and optical property changes, that is, by first isomerizing to its true acid form [equation (38)]. Pseudoacid strength could not be directly related to salt-forming tendency and the existence of an acid in pseudo form did not necessarily imply weakness; for example, Hantzsch found that hydrogen halides existed as true acids in ethanol, but that nitric acid, a stronger acid, existed as a pseudoacid isomer (NO_2OH) in the same solvent.[28]

An important result of Hantzsch's work was the discovery that the existence of an acid as pseudo or true (or as both isomers in dynamic equilibrium with each other) under a particular set of conditions depended on both the acid and the solvent involved. Early investigations utilized absorption spectrophotometry to experimentally distinguish between true and pseudoacids. Hantzsch compared the absorption spectra of carboxylic acids in various solvents to the spectra of their respective carboxylate salts and their esters in the same solvents.[26,27] If the acid spectrum resembled the spectrum of its (electrolyte) salt, it existed as a true acid in the solvent of interest; if the acid spectrum was close to the (nonelectrolyte) ester spectrum, then existence of the pseudoacid isomer was established. For example, trichloroacetic acid gave a true acid spectrum in water and a pseudoacid spectrum in diethyl ether:

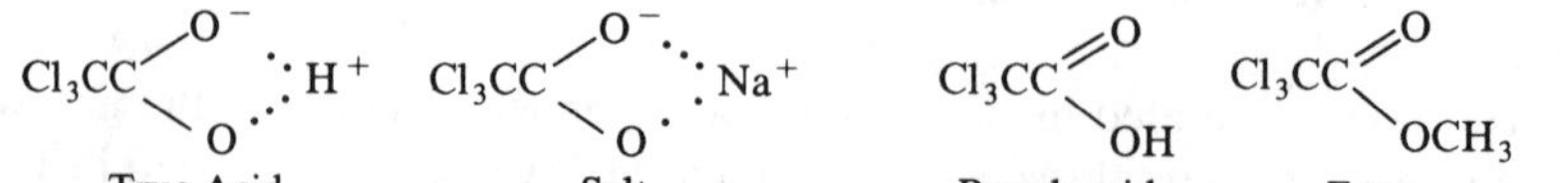

$Cl_3CC(O^-)(O) \cdots H^+$ — True Acid; $Cl_3CC(O^-)(O) \cdots Na^+$ — Salt (Ionic Bonding)

$Cl_3CC(=O)OH$ — Pseudoacid; $Cl_3CC(=O)OCH_3$ — Ester (Covalent bonding)

These experiments were later expanded to include inorganic oxyacids, which were found to exhibit pseudoacid behavior as the acid concentration approached 100%:

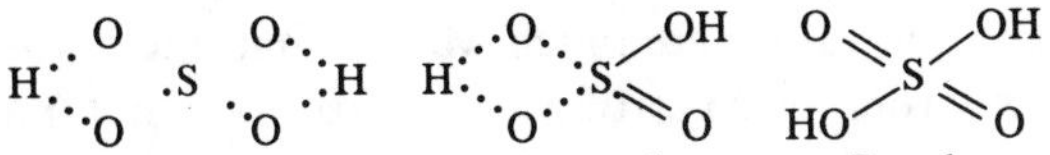

$H{:}O_2S{:}O_2{:}H$	$H{:}O_2{:}SO(OH)$	$O_2S(OH)(OH)$	$NO_3\cdots H$	NO_2OH
True	Semipseudo	Pseudo	True	Pseudo
Sulfuric acid			Nitric acid	

Even the hydrogen halides behaved as pseudoacids under certain circumstances. The absorption spectrum of pure hydrogen bromide differed from the bromide salt-like spectra of HBr in water, ether, and ethanol, and this was interpreted by Hantzsch as evidence of a pseudoacid isomer of HBr.[28]

Although the spectrophotometric measurements distinguished between true and pseudoacids, they contained no hint of the results of catalysis studies by Hantzsch.[28] Sugar inversion and diazoacetic ester decomposition, two processes commonly employed for measuring acid strength via catalytic activity, served as reactions for which further tests of the true and pseudoacid hypothesis could be carried out. Since it was believed that acid catalysis depended on the concentration of free or solvated protons,[16–19] it was not surprising to find little or no catalytic power in acid–solvent combinations with pseudoacid absorption spectra, for example, HCl in diethyl ether. However, true acids were much stronger catalysts in nonionizing solvents than in ionizing solvents, in direct opposition to the prevailing belief. It was also noted that the difference in catalyzing power between two acids did not remain constant with a change of solvent, indicating that the nature of the solvent, as well as the isomer present, also played a role in determining acid strength.

Based on these experiments,[26–30] Hantzsch arrived at a series of revolutionary conclusions concerning the general nature of acids and solvent effects on acidity. He proposed that the actual carriers of acid properties be undissociated true acids instead of hydrogen ions. The catalysis experiments had demonstrated less acid character for the ionized than for the undissociated acids.

Reasoning by analogy to the formation of weakly acidic ammonium salts from acids and ammonia, Hantzsch equated the ionization of an acid in an ionizing solvent with salt formation. The classical aqueous ionization concept was supplanted by the reaction of an acid with a base (water) to form a weaker, acidic hydronium salt, just as acids reacted with aqueous or pure ammonia to form analogous, weakly acidic ammonium salts. The replacement of ionization in a solvent by salt formation led Hantzsch to make some revisions in his theory. Both true and pseudoacids were considered to dissociate directly into salts in an ionizing solvent, the latter no longer requiring the intermediate formation of a true acid isomer.

Strong acids were particularly affected by ionizing solvents with basic properties, since their hydronium salts were much weaker than the undissociated acids, whereas the hydronium salts of weak acids were not much weaker than the acids themselves. Increasing the basic nature on the part of the solvent often blurred differences in acid strength, especially between strong acids. For example, HCl, H_2SO_4, HNO_3, and $HClO_4$ all seemed equally strong in water, because each of these acids were strong enough to react completely with water to produce equally weak hydronium salts, although the inherent strengths of these acids differed, that is, water "leveled" the strengths of all four acids. Hantzsch's leveling effect explained how variations in solvent basicity affected acid strength. Increased solvent basicity meant greater leveling and a decrease in observed differences in acid strength; decreased solvent basicity led to less leveling and increased strength differences, a fact clearly confirmed by Conant and Hall,[31,32] who relied on the increased salt-forming tendencies ("superacidity") of perchloric and sulfuric acids in glacial acetic acid (compared to their acidities in water) to titrate bases too weak to be titrated in aqueous solution. Hammett[30] derived equations to illustrate the solvent dependence of the leveling effect from the mass-action law.

In a later study Hantzsch found that the absorption spectrum of potassium nitrate in concentrated sulfuric acid did not resemble the expected pseudonitric acid spectrum of concentrated HNO_3 (NO_2OH) but, rather, the spectrum of dilute aqueous nitric acid [according to Hantzsch, hydronium nitrate, $(H_3O)(NO_3)$].[33,34] A salt-like spectrum was also obtained when perchloric acid was substituted for nitric acid. Hantzsch attributed these observations to the expected pseudonitric acid formation followed by a salt-formation reaction between the stronger acid H_2SO_4 and NO_2OH, where the latter played the role of a base, in much the same way as water behaved as a base in hydronium salt formation:

$$H_2SO_4 + NO_2OH \rightarrow [NO(OH)_2][HSO_4] \qquad (40)$$

$$2H_2SO_4 + NO_2OH \rightarrow [N(OH)_3][HSO_4]_2 \qquad (41)$$

Hantzsch called the salts formed from pseudoacids "pseudosalts" and claimed that, along with pseudoacids, they constituted an intermediate pseudoelectrolyte class in between the major classes of electrolytes and nonelectrolytes.[33] He used the concept of pseudo (or "acidium") salts to explain the limited conductivity of pure anhydrous acids[37] and mixtures of anhydrous acids, for example, $HClO_4$–HF, BF_3–HF, H_2SO_4– HCl.[35] For pure nitric acid, pseudosalt formation may be represented by

$$NO_2OH + NO_2OH \rightarrow [NO(OH)_2][NO_3] \quad (42)$$

$$2NO_2OH + NO_2OH \rightarrow [N(OH)_3][NO_3]_2 \quad (43)$$

These findings caused Hantzsch to drastically revise his theory of acids,[29,36,37] Ceasing to believe in the existence of true acids under any circumstances, he maintained that all undissociated acids were pseudoacids ionizing only via salt formation; that is, salts were the only true electrolytes and acids were nonelectrolytes that sometimes, for example, in aqueous solution, seemed to behave as electrolytes, but such behavior was actually a consequence of salt formation:

$$\text{Pseudoacid} + H_2O \rightarrow \text{Hydronium salt}$$

$$\text{Hydronium salt} \rightarrow \text{Dissociated salt (ions)} \quad (44)$$

The idea that all pure acids were pseudoacids was severely criticized and widely challenged,[8] but one side effect of the postulation of the existence of pseudosalts was to emphasize the relative nature of acidity and the generally amphoteric nature of substances, even of those commonly believed to possess only acid properties, for example, HCl and HNO_3.

In summary, Hantzsch's pseudoacid concept found little acceptance among chemists subsequent to his original experiments with nitro-organics. Even these reactions are now believed to proceed slowly because of electronic rather than atomic rearrangements within molecules.[6,8] Opponents of the theory also cited the spectral shifts upon which Hantzsch based many of his conclusions as too small from which to draw any significant conclusions. However, many of the ideas that emerged as adjuncts of the pseudoacid theory, for example, acidity of undissociated acids, ionizing and leveling solvent effects, basicity of water, and general amphoterism of substances, survived and attained importance upon incorporation into other acid–base concepts.[30]

REFERENCES

1. Day, Jr., M. C., and Selbin, J., "Theoretical Inorganic Chemistry," 2nd ed., Reinhold, New York, 1969.
2. Bjerrum, J., *Naturwissenschaften*, **38**, 461 (1940).
3. Hall, N. F., *J. Chem. Educ.*, **17**, 124 (1940).

4. Luder, W. F., *ibid.*, **25**, 555 (1948).
5. Kolthoff, I. M., "Treatise on Analytical Chemistry," Part I, Vol. I, Kolthoff, I. M., Elving, P. J., and Sandell, E. B., Eds., Wiley-Interscience, New York, 1959, Chap. 11.
6. Bell, R. P., "The Proton in Chemistry," 2nd ed., Cornell University Press, Ithaca, New York, 1973.
7. Kolthoff, I. M., *J. Phys. Chem.*, **48**, 51 (1944).
8. Bell, R. P., *Q. Rev. Chem. Soc.*, **1**, 113 (1947).
9. Franklin, E. C., *J. Am. Chem. Soc.*, **27**, 820 (1905).
10. Gillespie, R., "Proton Transfer Reactions," Caldin, E., and Gold, V., Eds., Chapman and Hall, London, 1975, Chap. 1.
11. Kahlenberg, L., *J. Phys. Chem.*, **6**, 1 (1902).
12. Folin, O., and Flanders, F. F., *J. Am. Chem. Soc.*, **34**, 774 (1912).
13. Flürscheim, B., *J. Chem. Soc.*, **95**, 718 (1909).
14. Collie, J. N., and Tickle, T., *ibid.*, **75**, 710 (1899).
15. Meyer, K. H., *Ber.*, **41**, 2568 (1908).
16. Goldschmidt, H., *ibid.*, **28**, 3218 (1895).
17. Goldschmidt, H., *ibid.*, **29**, 2208 (1896).
18. Fitzgerald, E., and Lapworth, A., *J. Chem. Soc.*, **93**, 2163 (1908).
19. Lapworth, A., *ibid.*, **93**, 2187 (1908).
20. Lapworth, A., *ibid.*, **107**, 857 (1915).
21. Vorlander, O., *J. Prakt. Chem.*, **87**, 84 (1913).
22. Pfeiffer, P., *Ber.*, **39**, 1864 (1906).
23. Pfeiffer, P., *ibid.*, **40**, 4036 (1907).
24. Werner, A., *ibid.*, **40**, 4133 (1907).
25. Hantzsch, A., *ibid.*, **32**, 575 (1899).
26. Hantzsch, A., *Z. Elektrochem.*, **24**, 201 (1918).
27. Hantzsch, A., *ibid.*, **29**, 221 (1923).
28. Hantzsch, A., *Ber.*, **58**, 612 (1925).
29. Hantzsch, A., *ibid.*, **60**, 1933 (1927).
30. Hammett, L. P., *J. Am. Chem. Soc.*, **50**, 2666 (1928).
31. Hall, N. F., and Conant, J. B., *ibid.*, **49**, 3047 (1927).
32. Conant, J. B., and Hall, N. F., *ibid.*, **49**, 3062 (1927).
33. Hantzsch, A., *Ber.*, **58**, 941 (1925).
34. Hantzsch, A., and Berger, K., *ibid.*, **61**, 1328 (1928).
35. Hantzsch, A., *ibid.*, **63**, 1789 (1930).
36. Hantzsch, A., *Z. Phys. Chem.*, **134**, 406 (1928).
37. Hantzsch, A., and Dungen, F., *ibid.*, **134**, 413 (1928).

CHAPTER

2

BRØNSTED–LOWRY THEORY

One might infer from most chemistry texts that the Brønsted–Lowry theory appeared suddenly following a 35-year period during which the acid–base concept was completely dominated by the ideas of Arrhenius. In fact, the proton-transfer theory incorporated and unified principles developed previously by Goldschmidt, Lapworth, Pfeiffer, and Werner (see Chapter 1), as well as those developed by Hantzsch simultaneously with the emergence of the Brønsted–Lowry approach in the 1920s.

2.1 ACID–BASE DEFINITIONS

The electrolytic dissociation theory attributed special character to the hydrogen ion because of its singular properties. The free proton was the *smallest, most mobile ion and possessed a high charge density*. In addition, of all the known chemical species it alone contained no electrons. The defects of the Arrhenius acid–base theory evidenced the need for a more general acid–base concept but did not detract from the belief in the special nature of the hydrogen ion relative to chemical species in general and to acids and bases in particular.

Brønsted first published his conception of acids and bases in 1923,[1] defining an acid as a substance capable of giving up a proton. Except for the omission of water as a requisite solvent, this definition did not seem to differ greatly from that of Arrhenius. However, Brønsted rejected the classical notion of bases as hydroxides as being too narrow, and instead defined a base as a substance with a tendency to accept protons. The theoretical relationship between acids and bases is summarized in the following defining equation[1-17]:

$$\text{Acid} \rightleftharpoons \text{Base} + \text{H}^+ \tag{1}$$

Lowry[18,19] independently reached the same conclusion in the same year from his observation of the "uniqueness of hydrogen."* According to Lowry, most substances were easily classified into one of two major categories, according to the intramolecular forces among their constituent atoms. Complete electron transfer from one atom to another characterized ionic substances. Electrostatic forces were operative between the resultant ions and dissociation and altered neither the ions themselves nor the nature of the attractive forces between them:

$$\text{Na}^+\text{Cl}^- \rightarrow \text{Na}^+ + \text{Cl}^- \tag{2}$$

Electron sharing was responsible for bonding in covalent compounds, and bond cleavage could be either homolytic or heterolytic; in the latter case the forces between the dissociation products differed from those between the original atoms. However, the breakage process was conceived of as abrupt, rather than gradual, so that the nature of the covalent bond remained unchanged prior to the instant of actual separation, and no prediction was possible as to whether homolytic or heterolytic cleavage would occur:

$$\text{Cl}^+ + \text{Cl}^- \leftarrow \text{Cl}:\text{Cl} \rightarrow 2\text{Cl}\cdot \tag{3}$$

Acids were exceptional when considered from this viewpoint, because they behaved as members of neither group, existing as covalent compounds when undissociated but invariably dissociating into ions:

$$\text{H}:\text{Cl} \rightarrow \text{H}^+ + \text{Cl}^- \tag{4}$$

Lowry attributed these observations to what he termed the "dualistic nature of hydrogen," that is, the unique ability of this element to assume a stable electronic configuration with either two or no electrons, for example, covalently as H:Cl and also ionically as H^+. A comparison of carboxylic

*Bell[39] has stated that Lowry deserves no credit as an originator of the protonic theory because of his failure to include explicit acid–base definitions in his original paper,[18] although many of his ideas implicitly lead to the Brønsted acid (but not base) definition. However, it is not the purpose of this work to apportion credit for the formulation of the protonic theory.

acids with carboxylate salts and esters, similar to Hantzsch's true–pseudoacid comparison, serves to illustrate this dualism[18]:

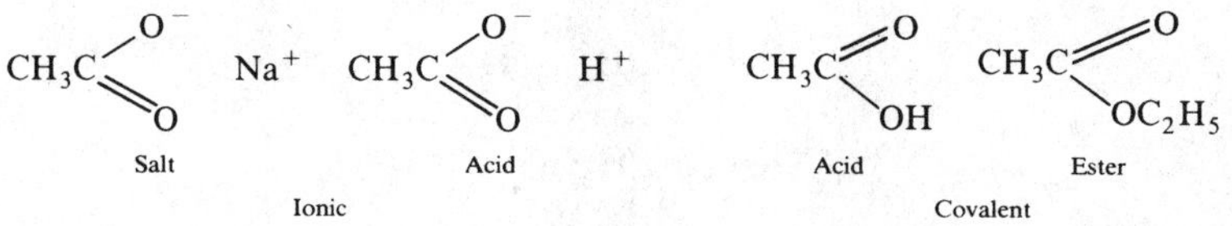

$$\underset{\text{Salt}}{CH_3C{\scriptstyle \begin{smallmatrix} O^- \\ O \end{smallmatrix}}\;Na^+} \quad \underset{\text{Acid}}{CH_3C{\scriptstyle \begin{smallmatrix} O^- \\ O \end{smallmatrix}}\;H^+} \qquad\qquad \underset{\text{Acid}}{CH_3C{\scriptstyle \begin{smallmatrix} O \\ OH \end{smallmatrix}}} \quad \underset{\text{Ester}}{CH_3C{\scriptstyle \begin{smallmatrix} O \\ OC_2H_5 \end{smallmatrix}}}$$

Ionic Covalent

Lowry envisioned the process by which a hydrogen bond changed from covalent to ionic as a gradual one during which ionic character increased with increasing interatomic distance:

$$H{:}Cl \rightarrow H^{\delta+} \cdots Cl^{\delta-} \rightarrow H^+ + Cl^- \tag{5}$$

where $\delta+$ and $\delta-$ represent partial positive and negative charges, respectively. Therein lay the essence of hydrogen's uniqueness.

An acid could not manifest its acidity (give up a proton) unless a species capable of accepting the proton was present, because, as had already been established, protons were not capable of independent existence. Lowry therefore also defined bases as proton acceptors.

Coincidentally, in 1923 Lewis[20] also proposed the proton donor–acceptor acid–base definitions but regarded them as too narrow and sought a more extensive theory.

The proton-transfer concept clearly expresses the reciprocal natures of acids and bases and bears a closer relation to their long-established fundamental "oppositeness" than the classical definition, which arbitrarily declares that hydrogen and hydroxide ions are "opposites."[1] Under the expanded definition of bases, the hydroxide ion, no longer restricted to aqueous solutions inherent in the Arrhenius concept, loses its former unique position. Alkali hydroxides are no longer regarded as bases but, rather as salts containing the base OH^-, in the same way as alkali chlorides are salts containing the base Cl^-.[1,6,9]

Ions are also classified as acids and bases if they indicate a tendency to either lose or gain protons. Whereas the classical view limits acids and bases to uncharged species manifesting their respective properties through ionization, the Brønsted–Lowry theory permits the existence of cationic and anionic acids and bases[1,4,9,16,17,25]:

$$[Fe(H_2O)_6]^{3+} \rightleftharpoons [Fe(H_2O)_5(OH)]^{2+} + H^+ \tag{6}$$

$$[Fe(H_2O)_5(OH)]^{2+} \rightleftharpoons [Fe(H_2O)_4(OH_2)]^+ + H^+ \tag{7}$$

$$NH_4^+ \rightleftharpoons HN_3 + H^+ \tag{8}$$

$$H_3O^+ \rightleftharpoons H_2O + H^+ \tag{9}$$

$$H_2O \rightleftharpoons OH^- + H^+ \tag{10}$$

$$CH_3COOH \rightleftharpoons CH_3COO^- + H^+ \tag{11}$$

$$H_2CO_3 \rightleftharpoons HCO_3^- + H^+ \tag{12}$$

$$HCO_3^- \rightleftharpoons CO_3^{2-} + H^+ \tag{13}$$

Brønsted and Wynne-Jones[21] supported this aspect of the theory by demonstrating that acids of differing charge types catalyzed orthoester hydrolysis.

The classical requirement that acid–base properties appear only through ionization, previously challenged by Folin and Flanders,[22] was discarded by Brønsted on the basis of his own observations of acid–base reactions in benzene.[3,6] More evidence contradicting the ionization prerequisite was obtained by Kilpatrick and Rushton,[23] who studied the rate of hydrogen gas evolution during dissolution of metallic magnesium in aqueous acid solutions. The evolution rate depends, as expected, on the hydrogen (hydronium) ion concentration for strong acids, but for weak acids the rate varies linearly with the concentration of undissociated acid. Furthermore, buffering the weak acid solutions, that is, decreasing the degree of ionization, raised the evolution rate. Therefore proton donors in general, rather than particular protonated species, are the true carriers of acid properties. The distinctive acid property is the availability of a proton instead of the actual donation process, allowing Brønsted to unequivocally regard, for example, HCl as an acid in benzene solution.

An acid–base pair related by equation (1), that is, two chemical species differing from each other by a proton, is referred to as a corresponding or conjugate acid–base pair,[1,4,15,16] for example ammonium ion and ammonia in equation (8). Every conjugate acid has only one conjugate base and vice versa, but this restriction does not prohibit a substance from serving as the acid of one conjugate pair and playing the role of a base in another conjugate pair. For example, water is a base in equation (9) and an acid in equation (10); similar behavior is ascribed to, among other things, hydroxopenta-aquoferric and bicarbonate ions [equations (6) and (7) and (12) and (13), respectively]. The Brønsted–Lowry theory is thus able to explain amphoterism,[1] the ability of a substance to act as either an acid or a base, simply by invoking equation (1). The Arrhenius concept had encountered difficulties in attempting to rationalize the observation that the same substance could apparently dissociate into either hydrogen or hydroxide ions (except if that substance was water).

Arrhenius defined neutrality as the condition of an aqueous solution containing equal hydrogen and hydroxide ion concentrations, based on the dissociation of water [equation (10)], but the Brønsted–Lowry view of hydroxide having no special significance as a base requires a reevaluation of the neutrality concept. The dissociation of other protic solvents, for example, methanol, is similar to that of water:

$$CH_3OH \rightleftharpoons CH_3O^- + H^+ \tag{14}$$

A methanolic solution containing equal hydrogen and methoxide ion concentrations is acidic according to the Arrhenius point of view, since no hydroxide ion is present, but according to Brønsted such a solution must be regarded as neutral, since methoxide ion is also a base. However, Brønsted did not suppose that hydroxide and methoxide are equally strong bases, and he therefore considered that there could be a difference between the acidity of a "neutral" aqueous solution and that of a "neutral" methanolic solution. Extension of this argument requires a different point of "neutrality" for every acid, since, in principle, any acid can serve as a solvent. The futility of trying to find a point of general neutrality emphasizes the arbitrary water-linked nature of neutrality heretofore accepted and this led Brønsted to declare the concept of neutrality as having no place in any acid–base theory.[1,4]

The invalidity of the neutrality concept will be explored in detail in Section 2.3.

2.2 ACID–BASE REACTIONS

The classical concept depicts an acid–base reaction as one producing water and a salt. Brønsted accepted neither product to be a necessary consequence of such a reaction and, as others before him, was able to show that acid–base reactions take place without the formation of water.[3,6] He also demonstrated that salt formation is not necessarily characteristic of only acid–base processes but that electron-transfer reactions also produce salts[3,17]:

$$Na + Cl \rightarrow Na^+Cl^- \tag{15}$$

$$(CH_3)_3N + CH_3I \rightarrow (CH_3)_4N^+I^- \tag{16}$$

A strict interpretation of the Brønsted–Lowry acid–base definitions can even be stretched to consider the reaction of, for example, aqueous HCl (a hydronium salt containing the acid H_3O^+) and NaOH (a salt containing the base OH^-) as salt destructive rather than salt forming,

$$\underset{\text{Salt}}{H_3O^+Cl^-} + \underset{\text{Salt}}{Na^+OH^-} \rightarrow \underset{\text{Salt}}{Na^+Cl^-} + H_2O \tag{17}$$

because there are two reactant salts but only one product salt.[6]

The Brønsted–Lowry conception of an acid–base reaction is derived from the inability of the proton to exist in the free state. An acid does not perform its characteristic proton-releasing function without the presence of an acceptor species, and a base cannot accept a proton unless an acid is present to donate one.[1,4,18] The fact that benzene lacks basic properties prevents HCl from manifesting acid properties in that solvent; that is, no ionization occurs, but the addition of a basic solute to such a solution results in an immediate acid–base reaction.[3,16]

Brønsted and Lowry concluded that an acid–base reaction involves the transfer of a proton from the acid to the base and can be represented as a double conjugate acid–base pair reaction[1–4,9,12,15–17]:

$$\begin{array}{c} \text{Acid}_1 \rightleftharpoons \text{Base}_1 + H^+ \\ \text{Base}_2 + H^+ \rightleftharpoons \text{Acid}_2 \\ \hline \text{Acid}_1 + \text{Base}_2 \rightleftharpoons \text{Base}_1 + \text{Acid}_2 \end{array} \tag{18}$$

The products of an acid–base reaction are simply the conjugates of the original acid and base.

Equation (18) embodies a practical definition of acids and bases as opposed to the theoretical one [equation (1)], the advantage of the former being that it implies proton transfer without requiring free proton formation, in agreement with experimental observation. Brønsted referred to equation (18) as "protolysis" and to the acids and bases involved therein as "protolytes".[5,16,24]

The protolysis scheme includes not only neutralization but also processes not previously regarded as acid–base reactions. The ionization process for an acid dissolved in a basic solvent is represented as protolysis, for example,

$$HCl + H_2O \rightleftharpoons Cl^- + H_3O^+ \tag{19}$$

but by no means are acid–base processes limited to aqueous media. Other proton-accepting solvents may be employed to demonstrate the acidity of

HCl; for example,

$$HCl + C_2H_5OH \rightleftharpoons Cl^- + C_2H_5OH_2^+ \tag{20}$$

Reversing the solute and solvent roles in equations (19) and (20) is still in accord with equation (18) and indicates that both water and ethanol behave as bases in liquid hydrogen chloride. The advantage of protolysis compared to the Arrhenius concept is clearly illustrated by considering the different perceptions of the basic reaction of ammonia in aqueous solution:

Arrhenius: $$H_2O + NH_3 \rightleftharpoons NH_4OH \rightleftharpoons OH^- + NH_4^+ \tag{21}$$

Brønsted: $$H_2O + NH_3 \rightleftharpoons OH^- + NH_4^+ \tag{22}$$

The restriction to aqueous solution associated with the Arrhenius concept precludes regarding a reaction between ammonia and another solvent, for example, ethanol, as a demonstration of the basic properties of the former:

$$C_2H_5OH + NH_3 \rightleftharpoons C_2H_5O^- + NH_4^+ \tag{23}$$

However, the protonic theory permits consideration of a reaction such as equation (23) as a manifestation of ammonia's basicity, since NH_3 is the recipient of a proton (and, by analogy to Arrhenius' requirement, the basic solvent anion is one of the reaction products). A redefinition of acids and bases in each solvent is not required by the expanded viewpoint of the Brønsted–Lowry theory, emphasizing the consideration of acidity and basicity as general phenomena independent of a particular solvent.

Hydrolyses (and solvolyses in general) also fit the protolysis scheme and they are no longer regarded as reactions of acidic and basic salts but are instead considered acid–base reactions in their own right:

$$NH_4^+ + H_2O \rightleftharpoons NH_3 + H_3O^+ \tag{24}$$

$$H_2O + CH_3COO^- \rightleftharpoons OH^- + CH_3COOH \tag{25}$$

Protolyses also include acid–base indicator reactions, since these are recognized as ordinary acid–base processes involving a conjugate pair in which the acid and base are of different color.

Thus the protolysis concept systematizes several classes of previously unrelated reactions in solution. It also extends the concept of acids and bases to include not only reactions taking place in solution but also proton-transfer processes in the gas phase and in melts.[26,27]

2.3 ACID–BASE STRENGTHS

Acidity and basicity are inversely varying functions, as defined by equation (1). No comparison between these two innately opposite effects is possible because, according to Brønsted,[1,4] no true point of neutrality exists. The difference between this viewpoint and the classical approach is illustrated by the diagrams of Figure 2.1. The Arrhenius theory depicts acidity and basicity as two symmetric, independent functions converging to zero at a point of neutrality. Comparisons of acidity are permitted only on the acid side of the neutral point, and comparisons of base strength only on the basic side. However, although the acidities of, for example, acetic acid and ammonia are not comparable under this scheme, it is possible to say that the acidity of acetic acid is equivalent to the basicity of ammonia. The latter statement is obviously predicated on the choice of water as a solvent, but experimental evidence indicates that it is not true in other media.[8] On the other hand, the Brønsted approach, discarding the aqueous reference state and the concept of neutrality, eliminates precisely this kind of ambiguity and pictures acidity and basicity as varying in opposite directions along a continuum. Acidities of substances can be compared, as can basicities, but acidity can never be compared to basicity.

Therefore the Brønsted–Lowry concept precludes neither a comparison of the acidities of HCl, NH_3, and OH^-, nor that of their basicities. The assertion of proton-donating tendencies, although small, for basic substances like hydroxide ion, or of proton-accepting abilities, also small, for

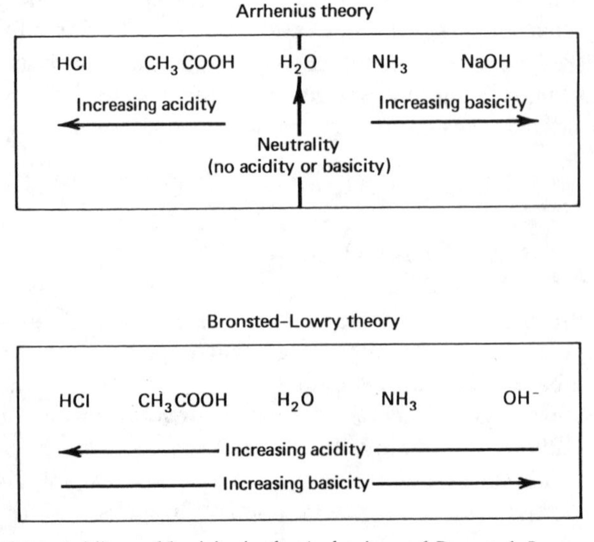

Figure 2.1. Acidity and basicity in the Arrhenius and Brønsted–Lowry theories.

acidic species such as HCl indicates the recognition of amphoterism as a general property, in agreement with Hantzsch's observations of mixtures of pure acids. Only the very strongest acids and bases possess virtually no properties of the opposite type.

Quantitative aspects of the protonic theory are derived from the application of the mass-action law to the theoretical [equation (1)] and practical [equation (18)] acid–base definitions. Brønsted defined an acidity constant K_A and a basicity constant K_B in terms of the concentration ratio of the participating conjugate acid–base pair and the hydrogen ion activity[4,5,14]:

$$K_A = \frac{C_B a_H}{C_A} \tag{26}$$

$$K_B = \frac{C_A}{C_B a_H} \tag{27}$$

Equations (26) and (27) represent, respectively, the tendency of an acid to lose a proton and that of a base to gain one, and these were regarded by Brønsted as measures of inherent acid–base strength, in principle independent of solvent nature.[9,10]

The reciprocal relationship of equations (26) and (27) is in accord with the general Brønsted conception of the link between acids and bases and consequently provides information concerning the strengths of conjugate acids and bases. For a given acid K_A is obviously the inverse of K_B for its conjugate base, leading to the conclusion that conjugate bases of strong acids are weak, and vice versa. It is not difficult to envision an acid that gives up a proton easily as one whose conjugate base makes no great effort to accept a proton, or an acid that gives up protons only with difficulty as one whose conjugate base has a large proton affinity. As one example of experimental confirmation of this assertion, consider the fact that chloride ion, the conjugate base of the strong acid HCl, does not hydrolyze in water, but acetate ion, the conjugate base of weak acetic acid, exhibits a marked tendency to pull protons off water molecules.

Absolute measurements of K_A and K_B are impossible, since acidity is not manifested in the absence of a proton acceptor, nor is basicity manifested without the presence of a proton donor.[3,6] Acidities are comparable only with regard to a reference base, and basicities can only be measured if a reference acid is present. The extent of such reactions is determined by applying the mass-action law to protolysis [equation (18)][2,6]:

$$K = \frac{C_{B1} C_{A2}}{C_{A1} C_{B2}} = K_{A1} K_{B2} \tag{28}$$

The protolysis constant K is a measure not of intrinsic acid–base strength but, rather, of the extent of proton transfer between an acid and a base, one of which may be acting as a reference substance. As a consequence, only relative acidities and basicities are obtained from equation (28).

Equation (28) explains the strong shift of protolysis equilibria towards the formation of weaker acids and bases.[10] The stronger the reacting acid and base, the larger the protolysis constant and the greater the extent of reaction; conversely, the weaker acid and base in a protolysis reaction are characterized by a protolysis constant that is the reciprocal of K for the stronger acid–base pair, and therefore the former react to a relatively small extent. The elementary formulation of this idea is the statement that strong acids and bases react to form weak acids and bases.

ACIDITY AND BASICITY OF SOLVENT

Since ionization of an acid or base in a solvent is a protolysis process,[3,6,16,24] the extent of such dissociation depends on the acidity or basicity of both the solute and the solvent and may be treated quantitatively by equation (28). Protolysis constants in water are identical to the Arrhenius dissociation constants[3,4,8,14]:

$$K_{a(\mathrm{aq})} = K_A K_{B(\mathrm{H_2O})} = K_{a(\mathrm{cl})} \tag{29}$$

$$K_{b(\mathrm{aq})} = K_{A(\mathrm{H_2O})} K_B = K_{b(\mathrm{cl})} \tag{30}$$

where cl indicates the classical constant.

The protolysis concept permits extension of acid–base dissociation into nonaqueous solvents. The latter are generally classified according to their acidic and basic properties relative to water. Predominantly acidic solvents, for example, glacial acetic acid and anhydrous sulfuric acid, are "protogenic," whereas solvents with far greater basic than acidic properties, for example, ammonia, are "protophilic." Solvents exhibiting both types of behavior, where neither predominates, are "amphiprotic"; this category includes water and ethanol. Finally, solvents with no observable acid–base properties are referred to as "aprotic" or "inert," for example, benzene and chloroform.[24,28,29]

These solvent categories are utilized within the confines of the protolysis concept to correlate a great deal of superficially unrelated information concerning aqueous and nonaqueous acid–base chemistry. Protolysis constants in a given solvent yield orders of acid and base strength in that solvent.[9] If one reverses one's frame of reference and regards a solute acid or base as a standard, relative solvent acidities and basicities, as well as comparisons of solute acidity and basicity in different solvents, are obtained.

Leveling Effect and Differentiation As an example of the observations explained by the Brønsted–Lowry theory, consider the fact that acids seem to increase in strength with increasing solvent basicity if the degree of acid dissociation is taken as a measure of acid strength.[16] This trend, from the viewpoint of protolysis, is seen as the result of an increase in the protolysis constant for the dissociation of a given acid as K_B of the solvent increases. Thus acetic acid, a weak acid in aqueous solution, behaves like a strong acid and dissociates completely in liquid ammonia, a more basic solvent. Analogous reasoning explains basic dissociation in solvents of increasing acidity.

In principle, the order of acid or base strength is invariant regardless of the solvent, since the ratio of the protolysis constants of any two acids or bases in any solvent is identical to the said ratio in any other solvent and to the ratio of the acidity or basicity constants of the species under consideration. In practice, however, many protolysis constants exceed the minimum value required for essentially quantitative reaction. Under these conditions differentiation between acid strengths (in highly basic solvents) or basic strengths (in highly acidic solvents) is impossible; that is, Hantzsch's leveling effect appears. Hantzsch understood the effect qualitatively, but the Brønsted–Lowry theory provides a quantitative interpretation based on protolysis constants.

Consider the example of two acids, one of which ($K_{A1}=10^5$) is inherently stronger than the other ($K_{A2}=10^{-5}$). The difference in strength is obvious in a basic solvent of $K_B=10^5$ in which the former is completely dissociated ($K_{a_1(B)}=10^{10}$) but not the latter ($K_{a2(B)}=1$). In a solvent of much greater basicity ($K_{B'}=10^{15}$) both acids are quantitatively ionized ($K_{a1(B')}=10^{20}$, $K_{a2(B')}=10^{10}$) and are indistinguishable in terms of strength.

The leveling effect imposes limits on both acidity and basicity, depending on the nature of the solvent.[24,28,29] Protophilic solvents place an upper limit on the acidity of solutions by leveling solute acid strengths to the strength of the conjugate acid of the solvent or, as Bjerrum[30] called it, the "lyonium" ion. Acids stronger than the lyonium ion possess acidity constants of a magnitude sufficient to produce a very large protolysis constant in combination with the solvent K_B and undergo quantitative protolysis; that is, no acid stronger than the lyonium ion exists in a solvent. The aforementioned behavior of acetic acid in liquid ammonia is an example of such leveling.

Analogously, protogenic solvents limit the basicity of solutions by leveling solute base strengths to the strength of the conjugate base of the solvent, which Bjerrum called the "lyate" ion. Amphiprotic solvents exert leveling effects at both ends of the acidity–basicity spectrum, and inert solvents level neither acids nor bases.

Hammett[31,38] provided a mathematical treatment of the leveling effect on protolysis. A simplified version of Hammett's treatment is presented here. Acidity is expressed for an acid HA in a basic solvent S by either the acidity constant of the acid or the acidity constant of the lyonium ion SH^+ (assuming a constant solvent concentration):

$$a_H = \frac{K_{HA}C_{HA}}{C_A} = K_{SH}C_{SH} \tag{31}$$

Electroneutrality requires that $C_A = C_{SH}$. If C is defined as the analytical concentration of HA, then

$$C = C_{HA} + C_A = C_{HA} + C_{SH} \tag{32}$$

Squaring equation (31) and substituting equation (32) gives

$$\begin{aligned} a_H^2 &= \frac{K_{HA}K_{SH}C_{SH}C_{HA}}{C_A} \\ &= K_{HA}C_{HA}K_{SH} = K_{HA}K_{SH}(C - C_{SH}) \\ &= K_{HA}K_{SH}C - K_{HA}a_H \end{aligned} \tag{33}$$

Applying the quadratic formula to equation (33) gives a solution for a_H:

$$\begin{aligned} a_H &= \frac{-K_{HA} + (K_{HA}^2 + 4CK_{HA}K_{SH})^{1/2}}{2} \\ &= \frac{K_{HA}}{2}\left[-1 + \left(1 + \frac{4CK_{SH}}{K_{HA}}\right)^{1/2}\right] \end{aligned} \tag{34}$$

A weak acid dissolved in a weakly basic (differentiating) solvent does not undergo complete protolysis ($K_{SH}/K_{HA} \gg 1$) and in such cases equation (34) reduces to

$$a_H = (CK_{HA}K_{SH})^{1/2} \tag{35}$$

indicating the dependence of acidity on both the acid and the solvent. On the other hand, in a strongly basic (leveling) solvent protolysis of even a weak acid may be complete ($K_{SH}/K_{HA} \ll 1$) and equation (34) becomes [via the mathematical approximation $(1+x)^{1/2} \sim (1+x)$ as x approaches zero]

$$a_H = CK_{\mathrm{SH}} \tag{36}$$

where the acidity depends only on the solvent; that is, a solution of an acid in a leveling solvent behaves as if it were a solution of the lyonium ion.

Hammett's approach is also useful in determining acidity and basicity in differentiating solvents. Equation (34) predicts an increase in acidity with decreasing solvent basicity, in agreement with the results obtained by Hantsch. At this point the distinction between the Brønsted concept of acidity (the tendency to denote a proton) and the protolysis process must be emphasized to avoid confusing acidity, which is dependent only on K_A, with dissociation, which depends on the protolysis constant $(K_A K_B)$. For example, the high acidity constant of HCl reflects the fact that it is a strong acid, but aqueous HCl solutions reflect this strength only because the protolysis constant of the HCl–water reaction is high. Proton transfer may not be complete in a less basic solvent, but HCl is still a strong acid. It is therefore not contradictory to distinguish between increasing true acidity in the Brønsted sense and increasing apparent acidity, which really stems from increasing solvent basicity. Consequently, HCl is a stronger acid in glacial acetic acid than in water, even though it is more completely ionized in the latter. This is due to the reduced extent of protolysis in the more acidic medium, which allows for the presence of the strong acid HCl in its undissociated from, whereas in water the leveling effect permits the presence of only the weakly acidic hydronium ion. In addition, what protolysis there is in acetic acid results in the formation of a much stronger lyonium acid than hydronium ion. Considerations such as these prove to be valuable in choosing suitable solvents for titrations.[28]

Conant and Hall[9,14,32,33] explained the observed "superacidity" of perchloric and sulfuric acids in glacial acetic acid within this context, although they also regarded salt formation as important. Bases too weak to be titrated in aqueous solution proved to be amenable to titration in the more acidic solvent, since the latter does not compete with the solute bases for protons to the extent that water does. A different interpretation of superacidity considers the phenomenon from a different reference frame, namely, that the acidic solvent levels weak bases so that they behave as strong bases. According to this concept, the term "superacidity" is misleading. The concept of superacidity is, in either case, a quantitative reevaluation of Hantzsch's ideas and again underscores amphoterism as a general property; for example, even acids behave as bases towards stronger acids.[15,16]

Another fact explained by the leveling effect is the observation of a constant heat of neutralization for reactions of strong acids with strong bases in aqueous solution. Since strong acids are leveled to hydronium ion, and strong bases to hydroxide ion, in aqueous medium, these ions are the actual reacting acid and base, and the heat of neutralization is constant regardless of the identities of the strong acid and base[8]:

$$H_3O^+ + OH^- \rightleftharpoons 2H_2O \tag{37}$$

Autoprotolysis The reverse of equation (37) is an example of solvent self-ionization resulting in the formation of lyonium and lyate ions.[4,16] Self-ionization, or "autoprotolysis," can occur only in amphiprotic solvents, because only these solvents possess both proton-accepting and donating abilities. Solvent autoionization also permits the establishment of a finite acidity scale in a given solvent. Protophilic solvents level acidity, but basicity in such solvents is indeterminate, though large. The converse is true for protogenic solvents, and in inert solvents both acidity and basicity can reach very large, indeterminate values.

Autoprotolysis is quantitatively characterized by an autoprotolysis constant. These constants are often small compared to other protolysis constants, because amphiprotic solvents are weak acids as well as weak bases; for example, the acidity constant of water is less than that of hydronium ion, and hydroxide ion is a stronger base than water, a consequence of the fact that protolysis favors the formation of weak acids and bases from strong ones. Even in cases at the fringes of the amphiprotic solvent category, where either the acidic or basic tendency is relatively strong, autoprotolysis constants remain small because the opposing tendency is extremely weak, as explained by the Brønsted–Lowry theory reciprocal picture of acidity and basicity. The weak acidity and basicity of amphiprotic solvents is attested to by their low conductivity in the pure state; the postulation of simultaneous strong acid and strong base behavior for a pure solvent implies a high protolysis constant and a high degree of ionization that is not in accord with experimental observation. Autoprotolysis constants also depend on the dielectric constant of the solvent.

The presence of an amphiprotic solvent does not conceptually alter the inverse relationship between the members of a conjugate pair, but includes the autoprotolysis constant of the solvent in their relationship. Applying equations (29) and (30) to a conjugate acid–base pair in aqueous solutions,[5,13,25,28,30]

$$K_{a(\mathrm{aq})} = \frac{C_B C_{\mathrm{H_3O^+}}}{C_A} \tag{38}$$

$$K_{b(\mathrm{aq})} = \frac{C_A C_{\mathrm{OH^-}}}{C_B} \tag{39}$$

makes it clear that, since $K_A K_B = 1$ for a conjugate pair [equations (26) and (27)], the product of the aqueous protolysis constants for the members of a conjugate pair equals the autoprotolysis constant of water, K_W [equation (37)]:

$$K_{a(\text{aq})} K_{b(\text{aq})} = K_{A(\text{H}_2\text{O})} K_{B(\text{H}_2\text{O})}$$

$$= C_{\text{H}_3\text{O}^+} \quad C_{\text{OH}^-} = K_W \tag{40}$$

{Note that K_W relates two different conjugate pairs [equations (9) and (10)] and is thus not equal to unity.} Equation (40) and its rationalization are applicable to amphiprotic solvents in general.

ACID–BASE CHARGE TYPE

The Brønsted–Lowry acid–base definitions imply that at least one member of every conjugate acid–base pair must be an ion. Any factor differentiating between species of different charges consequently affects acidity and basicity.

Increasing positive charge increases acidity and decreases basicity from the viewpoint of simple Coulombic forces. Increasing negative charge has the opposite effect.[6,25] The resistance of an anion to the loss of a proton or the attraction of an anion for a proton may be easily visualized. Conversely, the resistance of a cation to the addition of a proton and the ease with which a cation rids itself of a proton are also easy to envision. Brønsted[2,34] found that the acidity of complex aquometal ions increases with increasing charge; for example, $[Co(H_2O)(NH_3)_5]^{3+}$ is a stronger acid than $[Co(H_2O)(NH_3)_4Cl]^{2+}$. Comparisons between complexes of equal charge revealed that those containing a central metal atom in a higher oxidation state are also stronger acids; for example $[Pt(NH_3)_5Cl]^{3+}$ is more acidic than $[Co(NH_3)_6]^{3+}$.

Lowry[35] postulated intramolecular ionization to explain relative acid strength, mixing ionic and covalent bonding to produce structures like

HO–S^{2+}(–O^-)(–O^-)–OH and HO–S^{+}(–O^-)–OH

and attributing the greater acidity of sulfuric relative to sulfurous acid to greater repulsion of protons by divalent sulfur. Intramolecular ionization also provides an explanation for the increasing acidity observed along a row of the periodic table[6]; for example, an order of acidity such as

$$CH_4 < NH_3 < OH_2 < FH$$

may be regarded as the result of decreasing electrostatic attraction between protons and C^{4-}, N^{3-}, O^{2-}, and F^{-}, respectively. The relative strengths of the conjugate bases of these substances supports this idea: methide ion is extremely strong; amide ion is also very strong; oxide ion is strong; and fluoride ion is moderately strong.

Lowry[36] also advanced a different hypothesis to account for the relative acidities of such a series. Increasing nuclear charge exerts an increasing Coulombic attraction on orbiting electrons and contracts their orbitals, pulling the protons closer to the nucleus. The combination of increased nuclear charge with reduced internuclear distance increases the internuclear repulsion, causing a rise in acidity along the series from methane to hydrogen fluoride. Moreover, these effects are not limited to atoms directly attached to protons but may be transmitted through a chain of atoms; for example, chloracetic acid is stronger than acetic acid. Lowry invented the terms "acylous" and "basylous" to describe orbital-contracting (acidity-increasing) and orbital-expanding (basicity-increasing) atoms, respectively, though it is clear that what he actually observed was the influence of electronegativity.[34]

Brønsted defined acidity, basicity, and protolysis constants [equations (26)–(28)] in terms of concentrations of all participating species except the proton. These definitions are not fully in accord with the strict thermodynamic view of equilibrium, where activity replaces concentration. The activity a_i of a species i is related to its concentration by an activity coefficient γ_i:

$$a_i = \gamma_i C_i \tag{41}$$

The equilibrium constants defined by Brønsted are related to their thermodynamic counterparts K'_A, K'_B, and K' by terms composed of the activity coefficients of the participating species[4,5]:

$$K_A = \frac{a_B \gamma_A a_H}{a_A \gamma_B} = \frac{K'_A \gamma_A}{\gamma_B} \tag{42}$$

$$K_B = \frac{a_A \gamma_B}{a_B \gamma_A a_H} = \frac{K'_B \gamma_B}{\gamma_A} \tag{43}$$

$$K = K_{A1} K_{B2} = \frac{K'_{A1} K'_{B2} \gamma_{A1} \gamma_{B2}}{\gamma_{B1} \gamma_{A2}}$$

$$= \frac{K' \gamma_{A1} \gamma_{B2}}{\gamma_{B1} \gamma_{A2}} \tag{44}$$

In principle, the thermodynamic constants are more useful than those defined by Brønsted because they are independent of concentration effects, but the latter are advantageous in practice, since concentrations are relatively easy to determine. On the other hand, the multiple forms of the Debye–Huckel equation[37] attest to the fact that there is no universal agreement over the extent to which parameters relevant to the calculation of activity coefficients should be included in the said calculations. Furthermore, available activity data is limited to water and a handful of other solvents, rendering thermodynamic constants incalculable and useless in other instances.[6] Hammett[28,29,38] also realized that practical measures of acidity must be based on concentrations and defined "acidity functions" H_- and H_0 for acidic (HIn) and basic (In) indicators, respectively, in a variety of media accordingly:

$$\begin{aligned} pK_{\mathrm{HIn}} &= -\log\left(\frac{a_H \gamma_{\mathrm{In}^-}}{\gamma_{\mathrm{HIn}}}\right) - \log\left(\frac{C_{\mathrm{In}^-}}{C_{\mathrm{HIn}}}\right) \\ &= \mathrm{H}_- - \log\left(\frac{C_{\mathrm{In}^-}}{C_{\mathrm{HIn}}}\right) \end{aligned} \tag{45}$$

$$\begin{aligned} pK_{\mathrm{HIn}^+} &= -\log\left(\frac{a_H \gamma_{\mathrm{In}}}{\gamma_{\mathrm{HIn}^+}}\right) - \log\left(\frac{C_{\mathrm{In}}}{C_{\mathrm{HIn}^+}}\right) \\ &= H_0 - \log\left(\frac{C_{\mathrm{In}}}{C_{\mathrm{HIn}^+}}\right) \end{aligned} \tag{46}$$

Experimental factors capable of differentiating between species of varying charge affect acid–base strengths of protolytes of different charge types to different extents. Introduction of an inert electrolyte ("salt effect") and solvent or "medium" effects (dielectric and solvation effects) influence acidity and basicity by altering the activity coefficient ratios of equations (42)–(44) and consequently changing K_A, K_B, and K. Quantitative predictions of such changes are difficult to correlate with experimental observations for two reasons: both salt and medium effects involve some degree of approximation and the latter often result from the superimposition of several factors that are not easily separated. Nevertheless, separate consideration of salt and medium effects often proves to be valuable in predicting acid–base strength variations for protolytes of different charge types.

Salt Effects The thermodynamic acidity, basicity, and protolysis constants K'_A, K'_B, and K', respectively [equations (42)–(44)], may be written as products of concentration and activity coefficient terms [equation (41)]:

$$K'_A = \frac{a_H a_B}{a_A} = \frac{C_H C_B}{C_A}\frac{\gamma_H \gamma_B}{\gamma_A} = K'_{A(C)}\frac{\gamma_H \gamma_B}{\gamma_A} \tag{47}$$

$$K'_B = \frac{a_A}{a_H a_B} = \frac{C_A}{C_H C_B}\frac{\gamma_A}{\gamma_H \gamma_B} = K'_{B(C)}\frac{\gamma_A}{\gamma_H \gamma_B} \tag{48}$$

$$K' = \frac{a_{B1} a_{A2}}{a_{A1} a_{B2}} = \frac{C_{B1} C_{A2}}{C_{A1} C_{B2}}\frac{\gamma_{B1}\gamma_{A2}}{\gamma_{A1}\gamma_{B2}}$$

$$= K'_{(C)}\frac{\gamma_{B1}\gamma_{A2}}{\gamma_{A1}\gamma_{B2}} \tag{49}$$

where $K'_{A(C)}$, $K'_{B(C)}$, and $K'_{(C)}$ are concentration constants associated with the respective equilibria from which equations (47)–(49) are derived.

Activity coefficients of charged species in solution are often calculated from one of two forms of the Debye–Huckel equation.[25,29,37] The Debye–Huckel limiting law is used in dilute solutions ($<0.01\ M$),

$$\log \gamma_i = -Az_i^2 I^{1/2} \tag{50}$$

while in moderately concentrated solutions (0.01–$0.20\ M$) the Debye–Huckel expression gives better agreement with experiment:

$$\log \gamma_i = -\frac{Az_i^2 I^{1/2}}{1 + B\mathring{a} I^{1/2}} \tag{51}$$

where z_i is the charge on species i, $\mathring{a}$ is the ion-size parameter, A and B are the constants at a given temperature in a given solvent, and I is the ionic strength of the solution. Ionic strength depends on the concentrations and charges of all ions present:

$$I = \frac{1}{2}\sum_j C_j z_j^2 \tag{52}$$

The activity coefficient of an uncharged solute is also affected by ionic strength[25,37]:

$$\log \gamma_N = kI \tag{53}$$

where k is a constant.

Equations (51)–(53) are useful in the prediction of acidity changes caused by changing electrolyte concentrations. The activity of an uncharged species increases with increasing ionic strength, while that of a charged species decreases under the same conditions. One may visualize the latter effect as a consequence of increased interionic attraction at high concentrations. This attraction increases with increasing positive or negative charge, and the Debye–Huckel equations confirm that activity coefficients decrease more rapidly for ions of greater charge.

These considerations allow predictions concerning changes in acidity and basicity with variations in ionic strength for acids and bases of various charge types. Since K'_A and K'_B are measures of acidity and basicity, respectively, any change in acidity or basicity as a result of the salt effect must be reflected by a change in K'_A or K'_B.

Taking the negative logarithm of both sides of equation (47) yields

$$\mathrm{p}K'_A = \mathrm{p}K'_{A(C)} - \log\gamma_H - \log\gamma_B + \log\gamma_A \tag{54}$$

Addition of a neutral electrolyte, that is, a salt that neither increases nor decreases the concentrations of other species present, increases the ionic strength and alters the activities of all species present in a solution. The salt effect $\Delta \mathrm{p}K'_{A(s)}$ observed upon changing the ionic strength of a solution is therefore

$$\Delta \mathrm{p}K'_{A(s)} = \Delta(-\log\gamma_H - \log\gamma_B + \log\gamma_A) \tag{55}$$

since $\mathrm{p}K'_{A(C)}$ is constant. Equation (55) emphasizes the activity coefficient-dependent nature of the salt effect. An analogous derivation of $\Delta \mathrm{p}K'_{B(s)}$ from equation (48) yields

$$\begin{aligned} \Delta \mathrm{p}K'_{B(s)} &= \Delta(\log\gamma_H + \log\gamma_B - \log\gamma_A) \\ &= -\Delta \mathrm{p}K'_{A(s)} \end{aligned} \tag{56}$$

in accord with the Bronsted–Lowry view of the reciprocal natures of acidity and basicity.

$\Delta\log\gamma_H$ depends only on the magnitude of the change in ionic strength and is independent of the nature of the acid–base conjugate pair, but $\Delta\log\gamma_B$ and $\Delta\log\gamma_A$ depend on the charge type of the protolytes involved. This becomes clear if the activity coefficient of a species of charge z_i is expressed in terms of the activity coefficient γ_1 of an "average" univalent ion[29]:

$$\log\gamma_i = z_i^2 \log\gamma_1 \tag{57}$$

Since $z_B = z_A - 1$, substitution of equation (57) into equation (55) yields

$$\begin{aligned}\Delta pK'_{A(s)} &= -\Delta \log \gamma_H + (z_A^2 - z_B^2)\Delta \log \gamma_1 \\ &= -\Delta \log \gamma_H + (2z_A - 1)\Delta \log \gamma_1 \\ &= -\Delta \log \gamma_H + (2z_B + 1)\Delta \log \gamma_1 \end{aligned} \tag{58}$$

Equation (58) indicates that increasing ionic strength decreases the hydrogen ion activity and should consequently lower acidity. However, since $\Delta \log \gamma_1$ also decreases under these conditions, the magnitude of the decrease in the acidity of neutral and anionic acids exceeds that predicted by consideration of changes in hydrogen ion activity only. Cationic acids, on the other hand, tend to counteract the effect of $\Delta \log \gamma_H$, and therefore any decrease in acidity is likely to be smaller than expected. Equation (58) predicts increased acidity for $z_A > 1$, that is, some cationic acids may be expected to manifest stronger, rather than weaker, acidity with increasing ionic strength. Analogously, increasing ionic strength causes the expected increase in basicity of anionic bases to exceed that of neutral and cationic bases [equation (56)], and for highly charged cationic bases ($z_B > 1$) decreased basicity is predicted by an analogue of equation (58).

The influence of ionic strength on protolysis requires consideration of the salt effect upon both participating conjugate acid–base pairs. An expression for $\Delta pK'_{(s)}$ is obtained by combining equation (58) for one conjugate pair with an analogous equation for $\Delta pK'_{B(s)}$ for the second conjugate pair, or by direct derivation from equation (49):

$$\begin{aligned}\Delta pK'_{(s)} &= 2(z_{A1} - z_{A2})\Delta \log \gamma_1 \\ &= 2(z_{B1} - z_{B2})\Delta \log \gamma_1 \end{aligned} \tag{59}$$

Equation (59) indicates that protolyses involving two conjugate pairs of the same charge type are unaffected by the addition of a neutral electrolyte, since $\Delta pK'_{A(s)}$ of one conjugate pair is canceled by $\Delta pK'_{B(s)}$ of the other conjugate pair; that is, the total concentration of species of a given charge does not change in such a reaction and therefore no salt effect is predicted (although small changes in acidity and basicity due to solvation effects are often observed even in such cases).

Two specific examples that are widely applicable merit mention. First, considering only those protolyses involving acid or base dissociation in a solvent, equation (58) is reduced to

$$\Delta pK'_{a(s)} = 2(z_{A1} - 1)\Delta \log \gamma_1 = 2z_{B1}\Delta \log \gamma_1 \tag{60}$$

and, analogously,

$$\Delta pk'_{b(s)} = -2(z_{B2}+1)\Delta \log \gamma_1 = -2z_{A2}\Delta \log \gamma_1 \quad (61)$$

because a solvent behaves like a univalent cation acid–neutral base conjugate pair (represented by the symbol A^+B^0) towards solute acids and like a neutral acid–univalent anion base (A^0B^-) conjugate pair towards solute bases. Equations (60) and (61) are the two useful practical formulas for predicting salt effects in acidic and basic solutions.

The second example involves acid–base interactions with indicators, for which it was not previously recognized that the presence of inert electrolytes affects not only the activity coefficients of the protolytes of interest but also those of the indicator conjugate pair, making observations contingent on the indicator charge type. The confusing and seemingly contradictory results of acidity and basicity measurements with indicators prior to this realization had led many chemists to dismiss the technique as worthless, except under restricted circumstances.

As an example, consider the prediction of equation (58), that increasing ionic strength decreases the acidity of neutral and anionic acids but increases the acidity of cationic acids. An acidic indicator ($A2$=HIn and $z_{A2}=0$) confirms this for an acid of charge type A^+B^0 [equation (59)], but a basic indicator ($A2=HIn^+$ and $z_{A2}=1$) indicates no salt effect for the same acid; that is, the apparent acidity of the A^+B^0 acid is less than its true acidity. Likewise, the decreased acidity of A^0B^- acids under the same conditions is confirmed by basic indicators, but not by acidic indicators. Consequently, equations (58) and (59) are useful in predicting indicator errors caused by salt effects.[29]

Salt effect-related indicator errors may be generalized by considering the experimental criterion for indicator color, that is, the concentration ratio of the acidic and basic forms of the indicator in terms of indicator charge type:

Acidic indicator:
$$\frac{C_{\mathrm{HIn}}}{C_{\mathrm{In}^-}} = \frac{(\gamma_{\mathrm{In}^-})(a_H)}{(\gamma_{\mathrm{HIn}})(K_{\mathrm{HIn}})} \quad (62)$$

Basic indicator:
$$\frac{C_{\mathrm{HIn}^+}}{C_{\mathrm{In}}} = \frac{(\gamma_{\mathrm{In}})(a_H)}{(\gamma_{\mathrm{HIn}^+})(K_{\mathrm{HIn}^+})} \quad (63)$$

It may be concluded from equations (62) and (63) that increased ionic strength decreases the apparent acidity measured with acidic indicators and increases that measured with basic indicators [equations (50)–(53)]. Hammett[31] believed that the actual change in acidity as a result of the salt effect lay between the predictions of both indicator types.

Salt effects are difficult to predict under circumstances in which activity coefficients are not easily calculated. At high ionic strengths the Debye–Huckel equations are invalid. Even at moderate ionic strengths the ion-size parameter, å [equation (51)] is unknown in many cases. A high solute concentration approaches a change in solvent nature, superimposing medium effects onto salt effects. Finally, some types of molecules, for example, amino acids, proteins, and certain large indicator molecules, contain different acidic or basic sites that are far enough apart in the molecule to behave independently under some circumstances but not under others. These hybrid ions or "zwitterions" often exhibit a variation in apparent charge type with changing ionic strength that complicates the interpretation of observed salt effects. For example, an acid of charge type $A^{\mp}B^{-}$ can exhibit $A^{0}B^{-}$ behavior at low ionic strength but mixes $A^{+}B^{0}$ and $A^{-}B^{2-}$ behavior at high ionic strength because in the latter case its acidic sites behave independently of each other.[29]

Medium Effects The acidity or basicity of a solvent does not differentiate between acids or bases of different charge types, but a change of solvent introduces other factors that do. The acidity and basicity changes stemming from these factors are referred to as "medium" effects. A medium effect is formally defined as the work required to transfer one mole of a species from its standard state in water to its standard state in the solvent of interest[29] and it includes dielectric constant and solvation effects.

Activity coefficients of charged species are dependent on the dielectric constant ε because the constants A and B of the Debye–Huckel equations [equations (50) and (51)] incorporate it:

$$A \propto \varepsilon^{-3/2} \tag{64}$$

$$B \propto \varepsilon^{-1/2} \tag{65}$$

The majority of medium effects involve a decrease in dielectric constant, since most solvents have dielectric constants that are lower than that of water. The Debye–Huckel equation predicts that a lowered dielectric constant causes a corresponding decrease in ionic activity coefficients, the latter dropping more rapidly for ions of greater charge.

The medium effect $\Delta pK'_{A(m)}$ may be regarded as the change in acidity upon transfer of an acid from a reference medium R to the solvent of interest X:

$$\Delta pK'_{A(m)} = pK'_{A(X)} - pK'_{A(R)} \tag{66}$$

Since $pK'_{A(C)}$ [equation (54)] remains constant regardless of the solvent, an equation of a form similar to that of equation (55) is obtained:

$$\Delta pK'_{A(m)} = \Delta(-\log \gamma_H - \log \gamma_B + \log \gamma_A) \tag{67}$$

Therefore, following equations (57) and (58),

$$\Delta pK'_{A(m)} = -\Delta \log \gamma_H + (2z_A - 1)\Delta \log \gamma_1 \tag{68}$$

$\Delta \log \gamma_H$ depends only on the magnitude of the change in dielectric constant and is independent of the charge type of the acid–base conjugate pair. Equation (68) indicates that, as in the case of the salt effect, $\Delta \log \gamma_B$ and $\Delta \log \gamma_A$ are affected by the charges of the protolytes involved.

Equation (68) predicts that decreasing the dielectric constant decreases acidity, since the hydrogen ion activity is lowered, but in such a case $\Delta \log \gamma_1$ also decreases. Thus equation (68) predicts that the decrease in acidity for cationic acids is less than that expected from only considering the change in hydrogen ion activity and that neutral and anionic acids decrease in acidity with decreasing dielectric constant to a greater extent than expected. Anionic bases become stronger than expected, and cationic or neutral bases do not attain the level of strength predicted solely on the basis of $\Delta \log \gamma_H$.[4,5] The medium effect manifested by the decreasing dielectric constant therefore parallels the salt effect manifested by increasing ionic strength.

As in the case of the salt effect, the effect of a change of solvent on protolysis requires consideration of the charge types of both participating conjugate pairs. Equation (69), which is derived by analogy to equation (59), expresses the medium effect upon protolysis equilibria:

$$\Delta pK'_{(m)} = 2(z_{A1} - z_{A2})\Delta \log \gamma_1 \tag{69}$$

When applied to the special cases of acid and base dissociation in a solvent, equation (69) is reduced to [see equations (60) and (61)]

$$\Delta pK'_{a(m)} = 2(z_{A1} - 1)\Delta \log \gamma_1 \tag{70}$$

and

$$\Delta pK'_{b(m)} = -2(z_{B2} + 1)\Delta \log \gamma_1 \tag{71}$$

respectively, which are of practical use in determining medium effects. As in the case of the salt effect, equation (69) predicts that protolyses between conjugate pairs of the same charge type are not affected by a change of solvent. This is not always in agreement with experimental observation, since other effects, for example, changes in solvation, alter activity in ways that the foregoing equations cannot predict.

Medium effects also influence the reliability of indicator measurements and were responsible, along with salt effects, for incorrect and confusing assessments of acidity obtained by indicator methods in the past. One may consider the example of an A^+B^0 acid, whose increased acidity in a solvent of low dielectric constant is confirmed by measurements with acidic indicators but not by those with basic indicators, which show an apparent acidity that is less than the true acidity. Hammett's[31] treatment of the problem [equations (62) and (63)] showed that lowering the dielectric constant causes acidic indicators to show an apparent acidity that is less than the true acidity of a solution, while basic indicators show a higher apparent acidity; the true medium effect lies somewhere between these predictions.

A second approach to determining medium effects involves their derivation from the Born equation, which measures the work that is necessary to charge a species in a solvent. For one mole of a species i of charge z_i in a solvent of dielectric constant ε, this work is expressed by[28,29,40]

$$\Delta G_i^0 = -\frac{Ne^2 z_i^2}{2r_i}\left(1-\frac{1}{\varepsilon}\right) \tag{72}$$

where e is the electronic charge, N is Avogadro's number, and r_i is the radius of the species of interest, assuming a spherical shape. The first term of equation (72) represents the work required to discharge the species in vacuum, and the second term is the work necessary to charge the species in the solvent of interest; ΔG_i is therefore the solvation energy of species i.

In a protolysis reaction [equation (18)]

$$\begin{aligned} \Delta G^0 &= \Delta G_{A2}^0 - \Delta G_{B2}^0 + \Delta G_{B1}^0 - \Delta G_{A1}^0 \\ &= -2.3RT\log K' = 2.3RT\mathrm{p}K' \\ &= \frac{Ne^2}{2\varepsilon}\left(\frac{z_{A2}^2}{r_{A2}} - \frac{z_{B2}^2}{r_{B2}} + \frac{z_{B1}^2}{r_{B1}} - \frac{z_{A1}^2}{r_{A1}}\right) \end{aligned} \tag{73}$$

or

$$\Delta pK' = \frac{J}{\varepsilon}\left(\frac{z_{A2}^2}{r_{A2}} - \frac{z_{B2}^2}{r_{B2}} + \frac{z_{B1}^2}{r_{B1}} - \frac{z_{A1}^2}{r_{A1}}\right) \tag{74}$$

where the constant $J = Ne^2/4.6RT$.

At this point the Born equation makes an assumption that often leads to the difference between its predictions and experimentally observed medium effects. Solvation and nonelectrostatic effects are neglected; that is, the assumption is made that the radius of each species remains unchanged even with a change in solvent. This device allows the approximate determination of the medium effect for a change from a reference solvent R to the solvent of interest X:

$$\Delta pK'_{(m)} = J\left(\frac{z_{A2}^2}{r_{A2}} - \frac{z_{B2}^2}{r_{B2}} + \frac{z_{B1}^2}{r_{B1}} - \frac{z_{A1}^2}{r_{A1}}\right)\left(\frac{1}{\varepsilon_{(X)}} - \frac{1}{\varepsilon_{(R)}}\right) \tag{75}$$

If the reference solvent is the solvent of higher dielectric constant, the dielectric term of equation (75) is positive. Finally, to simplify equation (75) the assumption is made that the radii of the protolytes are not radically different. Since $z_A = z_B + 1$ for a conjugate acid–base pair, equation (71) becomes[28,29]

$$\Delta pK'_{(m)} = -\frac{2J(z_{A1} - z_{A2})}{r_{\text{avg}}}\left(\frac{1}{\varepsilon_{(X)}} - \frac{1}{\varepsilon_{(R)}}\right) \tag{76}$$

The validity of the last approximation is doubtful except for the members of a conjugate pair, but it does yield an equation of a form that qualitatively predicts medium effects in accord with the predictions of equation (69), which is derived directly from activity coefficients.

The simple Born treatment described above assumes that Born charging accounts for all of the electrostatic interactions between ions and the solvent. Although this is approximately true for large ions, the high charge density associated with small ions leads to the orientation and polarization of solvent molecules in the immediate vicinity of the ion. This results in the formation of a primary solvation shell around the ion, followed by a structure-broken region and then by the bulk solvent.[40] Consequently, the dielectric constant of the solvent at or near the surface of an ion is not identical to the bulk solvent dielectric constant, and thus equation (76) has only approximate, qualitative significance.

The foregoing discussions of medium effects omit the fact that there is a limit to how small a solvent dielectric constant can be before electrolytic dissociation is inhibited. Overall dissociation can be regarded as a two-step process: the first step is the actual ionization process, resulting in the formation of ion pairs, and the second step is the dissociation of these ion pairs. This is represented for acid dissociation in a solvent by[28]

$$HA + S \rightleftharpoons \underset{\text{Ion pair}}{SH^+A^-} \rightleftharpoons SH^+ + A^- \tag{77}$$

and for base dissociation by

$$B + SH \rightleftharpoons \underset{\text{Ion pair}}{BH^+S^-} \rightleftharpoons BH^+ + S^- \tag{78}$$

The latter step in each process depends on the dielectric constant, which influences the work required to separate the ions. The magnitude of this work is very small in solvents of high dielectric constant ($\varepsilon > 25$), and in such cases the ionization process is virtually identical to overall dissociation; medium effects in these solvents may be predicted by equations (69) and/or (76). However, electrolytic dissociation is repressed in media of very low dielectric constant because the work of ion separation is large; that is, equations (77) and (78) are effectively halted at the ion-pair stage. This is true for inert salts as well as protolytes, and therefore ionic strength in solvents of low dielectric constant may be indeterminate, consequently complicating the prediction of both salt and medium effects.

2.4 ACIDITY IN CONCENTRATED ACID SOLUTIONS. ACIDITY FUNCTIONS

Two techniques are commonly employed to measure the acidity or basicity of a solution (as distinguished from total acid or base content, which may be determined by titration): hydrogen ion activity is correlated with the potential of a suitable indicator electrode in potentiometric measurements; colorimetric or spectrophotometric measurements relate solution acidity or basicity to the concentration ratio of the protonated and unprotonated forms of an indicator, that is, a weak acid or base characterized by significant differences between the light absorption properties of its protonated and unprotonated forms, dissolved in the medium of interest.

Neither method is strictly applicable to measurements in concentrated acid or base solutions. The relatively scant knowledge of activity coefficient behavior and uncertain liquid junction potentials in concentrated solutions complicate the interpretation of potentiometric measurements. A method of extending the pH scale into concentrated acid solution by measuring the potentials of the oxidation states of several semiquinone dyes was proposed by Michaelis and Granick,[41] but the complexity of the underlying concepts appears to have prevented its wide application.

The indicators generally utilized in colorimetric and spectrophotometric measurements are useless at extremes of acidity or basicity because they are quantitatively protonated or deprotonated, respectively, under these conditions. Furthermore, Hammett[31,38] noted that indicator acidity or basicity measurements are unreliable, especially when the nature (or, more specifically, the dielectric constant) of the solvent is altered, because the concentration ratio of the protonated and unprotonated forms of the indicator depends on the activity coefficient ratio of these species. The variation in the latter ratio with changing medium differs for indicators of different charge types [equations (62) and (63)]. Hammett showed that lowering the dielectric constant of the medium often yields an apparent decrease in the acidity as measured with a basic indicator, whereas the same change in the medium causes an apparent increase in the acidity as measured with an acidic indicator. Neither measurement represents hydrogen ion activity, which is the theoretical measure of acidity. However, in order to provide a uniform convention for indicator acidity measurements in concentrated acid solutions, in 1932 Hammett and Deyrup[29,38,43,44] proposed that acidity be determined in such media with basic indicators, and they used a carefully selected set of indicators to develop an acidity scale in concentrated sulfuric acid–water mixtures.

The strength of a basic indicator B may be determined from the dissociation constant pK_{BH^+} of its conjugate acid;

$$pK_{BH^+} = -\log\left(\frac{a_H a_B}{a_{BH^+}}\right) = -\log\left(\frac{C_H C_B}{C_{BH^+}}\right) - \log\left(\frac{\gamma_H \gamma_B}{\gamma_{BH^+}}\right) \tag{79}$$

where a_i, C_i, and γ_i represent the activity, concentration, and activity coefficient, respectively, of species i, and where the activity coefficients follow the usual convention, that is, they become unity in infinitely dilute aqueous solution. The difference between the strengths of two bases B and C in a solution of given acidity may therefore be expressed by

$$pK_{BH^+} - pK_{CH^+} = -\log\left(\frac{C_B C_{CH^+}}{C_{BH^+} C_C}\right) - \log\frac{\gamma_B \gamma_{CH^+}}{\gamma_{BH^+} \gamma_C} \tag{80}$$

If the indicators are chosen so that their individual activity coefficient ratios γ_{BH^+}/γ_B and γ_{CH^+}/γ_C are identical in any given medium, the activity coefficient term of equation (80), then becomes zero. In such a case the difference in strength between B and C depends on the concentration term only and may thus be directly determined colorimetrically or spectrophotometrically. The requirement that indicators be selected so that their activity coefficient ratios cancel, also known as the "Hammett activity postulate," is an absolute necessity for valid comparisons in concentrated acid media by the indicator method, and this led Hammett to stipulate the following strict rules for the selection of suitable indicators.[38]

1. The indicator must be neutral or only slightly basic in pure water; that is, significant protonation should occur only in concentrated acid media.
2. The color of the indicator in anhydrous acid should be different from its color in water, and the color change should be reversible.
3. The charge type of the indicator must be well established. This requirement is essential for the Hammett activity postulate to be applicable. Activity coefficient ratios of indicators of different charge type may vary differently with changes in a medium, thus invalidating $pK_{BH^+} - pK_{CH^+}$ measurements.
4. The members of any set of basic indicators upon which an acidity scale is to be based must protonate via a single mechanism. Hammett's original indicator set was chosen so that the bases accepted protons via a "simple" mechanism, that is, the addition of a single proton at a well-defined site. Comparisons with species undergoing color changes due to the addition of molecular acid, hydronium ion, water, or multiple overlapping protons are not applicable to an acidity scale based on "simple" indicators. Mechanistic protonation differences are reflected in differences in activity coefficient ratios, which invalidate the Hammett activity postulate and any information obtained on the assumption that the postulate holds. Hammett classified indicators that exhibited $(+1, -1)$ electrolyte behavior in anyhydrous sulfuric acid (measured by cryoscopy or conductivity) as "simple."

Hammett and Deyrup colorimetrically measured indicator concentration (ionization) ratios C_{BH^+}/C_B, for selected organic basic indicators with proton-accepting tendencies ranging from moderately weak, that is, having some tendency to protonate in water, to extremely weak, that is, protonating only in concentrated or nearly anhydrous acid. Measurements were made for each indicator over a range of sulfuric acid–water mixtures, and when the logarithm of the ionization ratio was plotted against acid concentration, an approximately linear plot was obtained for each indicator. Hammett contended that any set of indicators for which the activity postulate holds may be recognized by the fact that the log ionization ratio-versus-acid concentration plots for the members of such a set yield a series of parallel lines,* except near the pure water and pure acid extremes of the acid concentration range, where medium effects interfere. Hammett also chose indicators with overlapping protonation ranges to permit the direct determination of indicator base strength differences [equation (80)] and was able to assign a pK_{BH^+} value to each indicator in his set, on a scale referenced to the aqueous standard state, despite the fact that ionization ratios were obtained in concentrated acid solution.[38,43,44] The latter was accomplished by including basic indicators of sufficient strength to be protonated in water, that is, bases of known pK_{BH^+}, in the indicator set, and by stepwise application of equation (80) at acid concentrations at which indicator protonation ranges overlap.

The inapplicability, due to activity coefficient considerations, of the conventional pH scale to concentrated acid solutions led Hammett[29,38,43,44] to develop a scale of acidity based on an acidity function H_0 defined as follows [cf. equation (79)]:

$$H_0 = -\log(a_H \gamma_H / \gamma_{BH^+}) = pK_{BH^+} - \log(C_{BH^+}/C_B)$$
$$= -\log h_0 \qquad (81)$$

where the subscript 0 denotes the charge of the neutral basic indicator,† and the relationship between H_0 and h_0 is identical to that between pH and hydrogen ion activity. $pK_{BH^+}, C_{BH^+}/C_B$, and consequently H_0 can be determined directly from experiment, thus avoiding the difficulties connected with accurate calculation of activity coefficients. It may be noted that the definition of H_0 becomes identical to that of pH in dilute aqueous solution and that, like pH, H_0 decreases with increasing acidity.

Hammett constructed a plot of H_0 versus acid concentration for 0–100% aqueous sulfuric acid solutions by combining ionization ratio measurements, pK_{BH^+} values, and the agreement between H_0 values determined by different indicators in acidity ranges in which individual indicator protonation

*The lines are "parallel" and "straight" within reasonable limits; the relative crudeness of Hammett's early colorimetric measurements made the observation of "parallelism" a subjective exercise, although the underlying principles were unaffected. The accuracy of ionization ratio measurements has improved considerably as spectrophotometry has replaced colorimetry.[44–47]

†Obviously, the determination of acidity by indicator methods is not limited to neutral basic indicators. Any set of acid–base indicators of a given charge type that obey the activity postulate and Hammett's other indicator selection criteria may be used to establish an acidity scale, using analogues of equations (79)–(81), for example, H_+ for $(+2, +1)$-type indicators, or H_- for $(0, -1)$ indicators.[42,44]

ranges overlap. Although indicator properties are used to determine H_0, the acidity function is indicator independent as long as the activity postulate holds. Thus the acidity function H_0 determined in aqueous H_2SO_4 is useful in other aqueous acid solutions; indicator pK_{BH^+} values measured in 0–70% aqueous percholoric acid agree with those measured in H_2SO_4, and H_0 for aqueous H_2SO_4, $HClO_4$, and HNO_3 are nearly identical (Hammett and others[42] have noted that H_0 for aqueous CF_3COOH solutions do not resemble those for aqueous mineral acids).

The principles behind the development of acidity scales from indicator measurements have found wide application owing to the relative simplicity of the techniques associated with ionization ratio determination. Many applications parallel those of the pH scale. Evaluation of the basicity of species not eligible for inclusion in Hammett's original indicator set and the investigation of differences in the behavior of various types of bases in strong acid solution have generated much scientific interest.

All Brønsted bases manifest basic properties by accepting a proton, but, as mentioned previously, not all do so via a single mechanism. The interplay of steric, inductive, resonance, solvation, hydrogen-bonding, and other effects makes for a wide variety of protonation modes. For example, Hammett chose indicator bases that protonate "simply," whereas triarylcarbinols expel a water molecule subsequent to protonation, leaving the triarylcarbonium ion as the analogue of the species BH^+. The activity of water, of marginal significance for "simple" Hammett indicators, becomes extremely important in the determination of triarylcarbinol basicity. The difference in protonation mechanisms makes it impossible for the determination of triarylcarbinol basicities on the Hammett H_0 scale to have any significance. However, an acidity function analogous to H_0 for bases undergoing triarylcarbinol-type ionization was established and utilized in the measurement of basicity of appropriate indicators and of medium acidity. Acidity scales were established for such diverse (in the sense of protonation mechanism) neutral basic species as aromatic hydrocarbons,[49] aliphatic and aromatic ethers,[50,51] arylcarbinols,[52–58] and diarylolefins.[59] Less common are scales utilizing indicators of different charge types; H_- was determined for cyanocarbon acids[60] and for aromatic amines,[61] and an H_+ acidity function has been reported.[62]

The potentially infinite number of possible acidity functions are not totally independent of each other, for, despite mechanistic differences, the net indicator reaction is fundamentally consistent. This fact was recognized by several investigators, who attempted to formulate mathematical relationships between the different acidity functions. One approach[63] utilizes a "standard" ion, that is, an ion chosen for its inertness towards reaction with the medium, for example, tetramethyl or tetraethylammonium ion, to relate several acidity functions. A second approach[64] expresses acidity functions as linear functions of each other, for example,

$$H_x = mH_0 + \text{const} \tag{82}$$

Hammett and Deyrup[38,65] demonstrated the applicability of the H_0 acidity function to nonaqueous solutions by comparing the results of pH and indicator measurements of solutions of H_2SO_4, benzenesulfonic acid, aniline, and sodium formate in water and in formic acid. While pH^{HCOOH} does not agree with pH^{H_2O}, pK_{BH^+} and H_0 values determined in the two solvents do agree. Therefore, acidity functions may be used to directly compare acidities in different solvents, unlike pH measurements, where intersolvent comparisons are complicated by the effect of a changing medium on activity coefficients. Nonaqueous acidity functions were established in many solvents, including glacial acetic acid,[66–69] dioxane,[70,71] ethanol,[70,71] acetone,[70,71] hydrocarbons,[72] and hydrazine.[73] Mixed alcohol–water media were employed to extend pK_{BH^+} determinations to weak bases that are insoluble in water.[74–76]

A number of investigations have focused on the influence of the solvent dielectric constant on the acidity function. Acidity functions often become indicator dependent in solvents with low dielectric constants; that is, the excellent correlation between acidity function values determined at a given acid concentration with different indicators in solvents of high dielectric constant no longer holds when the dielectric constant is sufficiently lowered.[44,69,72,76–78] This effect stems from the fact that ionization is fundamentally a two-step process, consisting of an ion-pair or aggregate formation step followed by an ion-separation or dissociation step; the latter is usually accompanied by solvation of the individual ions. In low-dielectric-constant media the second step is repressed and ionic solvation is not nearly as extensive as in high-dielectric solvents. The ionized indicator exists as part of an ion pair or aggregate rather than as an independent, solvated entity. The proximity of at least one other ion exerts varying degrees of influence on indicator species that would otherwise exhibit similar behavior, thus causing an acidity function based on these indicators to become indicator dependent.

Acidity scales have been established in unusual media by indicator methods, for example, heavy water.[79] Lewis and Bigeleisen[80] justified an acidity concept that was more general than that of Brønsted and Lowry (Chapter 4) by extending acidity measurements in sulfuric acid–water mixtures to show even greater acidity in $H_2SO_4-SO_3$ mixtures; this particular medium was also investigated by others.[81] Another strongly acidic medium in which an acidity scale was established for hydrogen acids by an indicator method is antimony pentafluoride.[82]

Temperature[83–87] and salt[44,83,88–92] effects on acidity functions, particularly H_0, were determined in studies paralleling those for the pH scale.

In addition to providing a great deal of data for very weak base and concentrated acid strengths, the acidity function concept has provided insight into activity coefficient behavior in concentrated acid solutions[57,93–96] and into the closely related influence of hydration on acidity.[39,86,95,97–100]

One consequence of the advent of the Brønsted–Lowry theory was increased interest in finding correlations between acidity and rates of acid-catalyzed reactions in an attempt to elucidate reaction mechanisms. Hammett found that a number of acid-catalyzed reactions with rates that were not successfully correlated with other measures of protonic acidity exhibited a linear relationship between rate and H_0,[43,83,101–103] and he concluded that the reactant in the rate-determining step of these reactions is a species formed by the protonation of a substrate S via a mechanism identical to that of the protonation of H_0 indicators:

$$S + H^+ \overset{k_1}{\rightleftarrows} SH^+ \tag{83}$$

$$SH^+ \overset{k_2}{\rightarrow} \text{Products or next intermediate} \tag{84}$$

Equation (84) represents the rate-determining step of the reaction mechanism. The linear relationship between the specific rate constant k of such a process and H_0 is expressed by

$$\log k + H_0 = \log k_2 \tag{85}$$

that is, a plot of $\log k$ versus H_0 is linear with a slope of unity. The mechanism of equations (83) and (84) and the rate law of equation (85) differ from their analogues in which undissociated acid, water, or hydronium ion are important. The former are catalyzed by H^+ only.

A wide variety of chemical reactions exhibit acidity–rate relationships of the type shown in equation (85).[46,102,103–105] Other processes exhibit a linear relationship between $\log k$ and H_0 with a slope other than unity.[106,107] The interplay of steric, hydrogen-bonding, inductive, resonance, solvation, and other effects differs for the substrates of such reactions from that characteristic of H_0 indicators. Properly selected indicators can be used to establish an acidity scale that, when plotted against $\log k$, yields a slope of unity [cf. equation (82)].

A more complex method of distinguishing between mechanisms was developed by Bunnett,[108–111] who plotted log $(k_\psi + H_0)$ against the activity of water (k_ψ is a pseudo-first-order rate constant) and grouped reactions with identical mechanisms according to the values of the slopes of these plots.

One noteworthy result of the application of acidity functions to kinetics is the fact that the rates of aromatic nitration reactions correlate well with the acidity function developed from triarylcarbinol indicators.[112,113] This suggests a rate-determining step very similar to carbinol ionization:

$$H^+ + HNO_3 \rightarrow NO_2^+ + H_2O \tag{86}$$

that is, protonation followed by the expulsion of a molecule of water. This view led to the rather interesting idea that charge delocalization is not as important to the formation of aromatic carbonium ions as was previously believed, since NO_2^+ is a small ion with limited charge-delocalizing ability.

Although acidity comparisons were facilitated by Hammett's approach, the effect of the constantly changing medium required for the establishment of an acidity scale does complicate acidity comparisons in a fundamental way. From the time he originally proposed the use of acidity functions, Hammett recognized that the absence of chemical side reactions does not preclude specific, unpredictable changes in the light absorption properties of indicators owing to changes in the medium.[38,44,114,115] These changes affect not only the absorptivity of a given species, but also the wavelength of its maximum absorption; the latter effect is referred to as "lateral displacement."[116] Errors introduced into pK_{BH^+} and H_0 determinations by these medium effects may invalidate the results of any applications of an acidity scale. Hammett and Deyrup[38] noted that the largest departures from parallelism in log (C_{BH^+}/C_B)-versus-acid concentration plots occur in the very dilute and very concentrated acid regions (where maximal activity coefficient changes are expected); H_0 also changes more rapidly with acid concentration at these extremes than in the intermediate concentration range. Hammett attributed these observations to large medium effects in nearly pure water or acid.

A number of attempts were made to mathematically correct for or experimentally eliminate the effects of a changing medium; for example, pure water or concentrated sulfuric acid;[38] measurements of absorptivities of completely protonated and deprotonated indicator forms in media of composition as close as possible to the medium of interest,[44] instead of in pure acid and pure water, respectively; and calculation methods.[116]

A different type of medium effect brought the existence of acidity functions in all media into question. It was argued that the dependence of the acidity function upon the indicators used to establish it in low-dielectric-constant media voids the acidity function concept entirely for such media because the Hammett activity postulate no longer holds,[29,76,117] although this contention was disputed as merely the result of a bad choice of indicators.[29,44]

Arnett and Mach[118,119] noted that the set of indicator bases utilized by Hammett to establish the original H_0 acidity function includes species of different organic functional groups, which necessarily differ in the combination of effects determining their mechanisms of protonation and consequently differ in the manner in which their activity coefficients vary with a change in medium; that is, Hammett's set of indicator bases does not really meet his own selection criteria. Arnett and Mach attributed the lack of parallelism in Hammett's log (C_{BH^+}/C_B)-versus-acid concentration plots to failure of the Hammett activity postulate for the indicator set, rather than the effect of a changing medium. The implication of severely restricted applications of the activity postulate imparts a fundamental weakness to the acidity function concept, since a separate acidity scale is required for each class of bases, but even more so because it becomes difficult to accurately classify given bases. The resulting confusion over the acidity function to be used in a particular instance is manifested in erroneous inferences drawn from the misapplication of acidity scales, for example, the elucidation of an incorrect reaction mechanism.

Arnett and co-workers,[120–123] having discovered a linear correlation between $\Delta H_{a,b}$, the heat of protonation of a weak base in a strong acid solution, and pK_{BH^+}, proposed a calorimetric method to compare weak base strengths in a strong acid solution as an alternative to the indicator method. The heat of solution, ΔH_S^{acid}, of the base of interest in a concentrated strong acid solution, for example, H_2SO_4 or HSO_3F, is measured and corrected for the energy associated with the separation of molecules by subtracting $\Delta H_S^{\text{inert}}$, the heat of solution, at infinite dilution, of the same base in an inert solvent, for example, CCl_4 or $Cl_2HCCHCl_2$:

$$\Delta H_{a,b} = \Delta H_S^{\text{acid}} - \Delta H_S^{\text{inert}} \tag{87}$$

Arnett noted several advantages of the calorimetric method:

1. Measurements are made in a single medium, eliminating the medium effect problem associated with the indicator method.
2. Protonation is complete, well defined, "clean" (no other reactions occur), instantaneous, and reversible. Cryoscopic, conductivity, NMR, and/or UV evidence to this effect was required of all bases included in the calorimetric basicity scale.
3. The method is characterized by accuracy and simplicity.

The only interference with the calorimetric approach occurs in cases where side reactions interfere with clean protonation, for example, aromatic sulfonation.

Arnett et al.[124–126] also developed an analogous calorimetric method for determining the strengths of very weak acids in strong basic media. The method utilizes the dimsylate ion, $DMSYL^-$, which is the lyate ion of dimethyl sulfoxide (DMSO). The heat of deprotonation, ΔH_D, of an acid in 0.1 M alkali $DMSYL^-$ solution in DMSO is the difference between its heats of solution, $\Delta H_S^{DMSYL^-}$ and ΔH_S^{DMSO}, in 0.1 M $DMSYL^-$ and DMSO, respectively:

$$HA + DMSYL^- \rightarrow DMSO + A^- \tag{88}$$

$$\Delta H_D = \Delta H_S^{DMSYL^-} - \Delta H_S^{DMSO} \tag{89}$$

The quantity of interest, however, is not ΔH_D, but, rather, ΔH_i, the heat of ionization, which occurs via the following reaction:

$$HA + DMSYL^- \rightarrow DMSOH^+ A^- \tag{90}$$

Subtraction of equation (88) from equation (90) indicates that ΔH_i may be obtained from ΔH_D if the heat of solvent autoprotolysis, ΔH_{auto}, is known:

$$2DMSO \rightleftharpoons DMSOH^+ + DMSYL^- \tag{91}$$

$$\Delta H_{auto} = \Delta H_i - \Delta H_D \tag{92}$$

ΔH_{auto} may be obtained by determining ΔH_D [equation (89)] for an acid of sufficient strength to be completely dissociated in DMSO, for example, HSO_3F.

2.5 ACIDITY AND BASICITY IN THE GAS PHASE. PROTON AFFINITY

The Brønsted–Lowry theory acknowledges the existence of acidity, basicity, and proton transfer in the absence of a solvent. Application of the protonic acid–base concept to the gas phase generated interest almost from the advent of the theory,[127] despite the lack of experimental methods for measuring acid–base strength in the gas phase at the time. What scarce data was available was obtained from calculations based on thermodynamic cycles.[127]

The importance of gas-phase acidity and basicity lies in the fact that the complicating effects of solvation are absent in the gas phase,[128] that is, only intrinsic acid–base properties are manifested under these conditions. Many supposedly "true" or anomalous orders of acid–base strength in solution, for example, the relative basicities of methylamines in aqueous solution, appear to be due to solvation effects when compared to their gas-phase strengths. A thorough understanding of gas-phase acidity and basicity permits a distinction between true and solvation-influenced orders of strength, indicates the magnitude of the effect of solvation on strength,[129] and may either confirm or invalidate interpretations of the relationship between structure and strength originally based on acid–base strengths in solution.

The conventional measure of gas-phase acidity or basicity is proton affinity, PA, defined for a species B of charge z as[130]

$$B^{z} + H^{+} \rightarrow BH^{z+1} \qquad PA(B^{z}) = -\Delta H \tag{93}$$

that is, the enthalpy required to remove a proton from BH^{z+1}. Increasing basic strength (or decreasing conjugate acid strength) is thus associated with increasing proton affinity.

Proton affinity may be broken down into several fundamental thermodynamic quantities.[131–140] PA is equivalent to the heterolytic bond dissociation energy $D(B^{-}\text{–}H^{+})$ for a (0, −1)-charge-type conjugate acid–base pair and may be regarded as a combination of the following thermodynamic quantities:

1. The homolytic BH bond dissociation energy D(B–H).
2. The electron affinity EA of the neutral radical B, which is a measure of the stability of B^{-}.
3. The ionization potential of the hydrogen atom, IP(H).

Thus

$$B\text{–}H \rightarrow B + H \qquad \Delta H = D(B\text{—}H) \tag{94}$$

$$B + e^{-} \rightarrow B^{-} \qquad \Delta H = -EA(B) \tag{95}$$

$$H \rightarrow H^{+} + e^{-} \qquad \Delta H = IP(H) \tag{96}$$

$$B\text{—}H \rightarrow B^{-} + H^{+}$$

$$\Delta H = PA(B^{-}) = D(B^{-}\text{–}H^{+}) = D(B\text{–}H) - EA(B) + IP(H) \tag{97}$$

Analogous treatment of the energetics of dissociation for a (1,0)-charge-type acid–base conjugate pair yields[141–145]

$$PA(B) = HA(B^+) + IP(H) - IP(B) \tag{98}$$

where $HA(B^+)$ represents the hydrogen affinity, that is, the ability of B^+ to latch onto an atom of hydrogen.

Since the ionization potential of the hydrogen atom remains constant (313.6 kcal/mole), the difference in proton affinity between two bases B_1^- and B_2^-, that is, the heat of proton transfer from B_1^- to B_2^-, may be expressed as

$$\begin{aligned} \Delta H = \Delta PA &= PA(B_1^-) - PA(B_2^-) \\ &= D(B_1{-}H) - EA(B_1) - D(B_2{-}H) + EA(B_2) \end{aligned} \tag{99}$$

Proton affinities may be reported in the literature either in absolute or relative (D–EA) terms, and it is therefore important to ascertain the manner in which a set of PAs is presented to avoid confusion or misinterpretation when comparing results from different sources.

The development of mass spectroscopy (MS) and ion cyclotron resonance (ICR) spectroscopy provided instrumental techniques applicable to the measurement of acidity and basicity in the gas phase. Absolute proton affinities were originally determined experimentally by the appearance potential mass spectrometric method.[146–148] The ionizing voltage, that is, the energy of the ionizing electron beam, is reduced to the minimum necessary to observe the protonated species of interest as a fragment ion in the mass spectrum. The heat of formation of the fragment ion, $\Delta H_f(BH^+)$, is obtained from this "appearance" potential, and the proton affinity of B is calculated from the heats of formation of the species in equation (93):

$$PA(B) = \Delta H_f(H^+) + \Delta H_f(B) - \Delta H_f(BH^+) \tag{100}$$

However, PAs obtained via the appearance potential method are inaccurate because excess internal excitation energy is included in the $\Delta H_f(BH^+)$ values calculated from appearance potentials. Furthermore, the technique is inapplicable to protonated molecules that do not occur as fragment ions.[148]

Haney and Franklin[148] attempted to correct appearance potential proton affinities for the excess kinetic energy by measuring the translational energy of the products of the following ion–molecule reaction, which takes place in an MS ionization chamber:

$$BH^+ + BH \rightarrow BH_2^+ + B \tag{101}$$

The heat of reaction of equation (101), ΔH^0, is estimated at five times the measured translational energy, and the PA of BH is calculated from

$$\mathrm{PA(BH)} = \Delta H_f(\mathrm{H^+}) + \Delta H_f(\mathrm{B}) - \Delta H_f(\mathrm{BH^+}) - \Delta \mathrm{H}^0 \qquad (102)$$

However, the correction is empirical and only approximate, and equations (101) and (102) exclude aprotic basic species from consideration.

Tal'rose[149,150] suggested that observation of the occurrence or nonoccurrence of proton transfer between two bases $B_1^{z_1}$ and $B_2^{z_2}$ be taken as an indication of their relative PAs; that is, for a given gas-phase protolysis reaction

$$\mathrm{HB}_1^{z_1+1} + \mathrm{B}_2^{z_2} \rightarrow \mathrm{HB}_2^{z_2+1} + \mathrm{B}_1^{z_1} \qquad (103)$$

the following three results in accordance with the Brønsted–Lowry theory are possible:

1. Reaction is observed in the forward but not in the reverse direction; equation (103) is therefore exothermic in the forward direction, and $B_2^{z_2}$ is a stronger base than $B_1^{z_1}$, that is, $\mathrm{PA}(B_2^{z_2}) > \mathrm{PA}(B_1^{z_1})$.
2. Reaction is observed in the reverse but not in the forward direction; equation (103) is endothermic as written, and the proton affinity of $B_1^{z_1}$ exceeds that of $B_2^{z_2}$.
3. Reaction is observed in both directions, that is, equation (103) is approximately thermoneutral and the proton affinities of the two bases are approximately equivalent.

Repeated application of this technique results in a scale of gas-phase acidity or basicity that is constructed in a stepwise manner.

Since the occurrence–nonoccurrence approach is founded upon protolysis reactions between two acid–base conjugate pairs instead of upon measurements involving only a single conjugate pair [cf. equation (93)], all PA data thus obtained are relative, that is, in terms of ΔPA [cf. equation (99)]. The absolute values often reported in the literature are actually relative values anchored to standards whose absolute PAs were previously determined by other methods. Two commonly employed standards are NH_3, whose PA was determined by the appearance potential method, and chloride ion, whose PA was determined by independent measurement of D(H–Cl) and EA(Cl).[151] If the proton affinities of the standards are found to be in error, the entire PA scale must be shifted to correct for the error, although relative proton affinities remain unchanged.

The occurrence–nonoccurrence approach is the method of choice in both MS and ICR. Although the data obtained is only relative, the uncertainties associated with absolute measurements are absent. Consequently, the absolute methods are no longer used except when technical improvements permit refined measurements for the PA standards.

Mass spectroscopy usually requires very low operating pressures to avoid deexcitation of the ions formed through collision or reaction with other molecules, and a good vacuum is essential for most MS work. However, the application of mass spectroscopy to gas-phase acidity and basicity measurements requires conditions that are favorable to chemical reaction, and thus relatively high operating pressures (0.1–10 torr) are used in this work.

Early high-pressure MS data were qualitative in nature.[130] Quantitative data were obtained by a number of investigators,[152–154] most notably by Kebarle and co-workers.[131–134, 155–169] Kebarle measured proton-transfer equilibrium constants for reactions involving either two $(0,-1)$-or two $(+1,0)$-charge-type acid–base conjugate pairs at high temperature ($\sim 600°K$) by measuring the steady-state abundances of the charged participants in the reaction at a known partial-pressure ratio of the uncharged species. The temperature dependence of the equilibrium constant is obtained from measurements at more than one temperature and is of the form

$$\ln K = \frac{-\Delta H}{RT} + \frac{\Delta S}{R} \qquad (104)$$

where K is the equilibrium constant and the other symbols have their usual thermodynamic significance. A Van't Hoff ($\ln K$-versus-$1/T$) plot yields the value of ΔH, which represents the net result of the competition between two bases for a proton, that is, the difference between their PAs. Stepwise application of this technique to determine unknown relative PAs from known ones yields a quantitative acidity or basicity scale in the gas phase.

There are two assumptions implicit in Kebarle's method of measuring gas-phase acidity and basicity: first, entropy effects [equation (104)] are assumed to be negligible; this view is not unreasonable if the ΔS associated with the deprotonation of one species is regarded as being approximately canceled by the ΔS associated with the protonation of the other species. The assumption of negligible entropy is experimentally verified not only for competing bases of similar structure, but even for proton-transfer reactions where large changes in entropy are predicted. ΔS is significant only for a small minority of proton-transfer processes.[157, 164]

The second assumption is that the linear relationship between $\ln K$ and $1/T$ found at the high temperatures required to maintain all constituents as gases holds at room temperature, which is approximately 300°K lower than the temperature at which measurements are made.

An accuracy of ±0.2 kcal/mole is claimed for PA values obtained from high-pressure MS, but this statement is valid only insofar as relative PAs are concerned; that is, only the differences between proton affinities are accurate to ±0.2 kcal/mole.

Bohme and co-workers[170] criticized the validity of the high-pressure MS technique, contending that thermodynamic equilibrium is not established under high-pressure MS measurement conditions and that protonated species are not thermalized by collisions prior to reaction. Consequently the equilibrium constants from which proton affinities are determined are in error. Bohme[171–175] utilized an alternative kinetic approach to gas-phase acidity and basicity, measuring the rate constants of forward and reverse protolysis equilibria in the flowing afterglow region of a mass spectrometer, where thousands of collisions thermalize the protonated reactants prior to proton transfer and allow reactions to proceed at 300°K. The equilibrium constant and the preferred direction of proton transfer are obtained from the forward and reverse rate constants k_1 and k_2, respectively[171–175]:

$$\frac{k_1}{k_2} = K = \frac{K_{A(g)}(B_1^{z_1})}{K_{A(g)}(B_2^{z_2})} \tag{105}$$

or

$$\log K = \Delta \mathrm{p}K_{A(g)} = \mathrm{p}K_{A(g)}(B_2^{z_2}) - \mathrm{p}K_{A(g)}(B_1^{z_1}) \tag{106}$$

where $K_{A(g)}(B_i^{z_i})$ represents the gas-phase dissociation constant of the conjugate acid of the base B_i of charge z_i.

Chemical ionization mass spectroscopy is another MS technique that was applied to gas-phase acidity and basicity determinations.[176]

Ion cyclotron resonance spectroscopy is the alternative to MS in the investigation of gas-phase phenomena. An ICR spectrometer operates at pressures that are sufficiently low and with ion residence times that are sufficiently long to allow extensive ion–molecule reaction. Ions are identified by a "single resonance" technique: ions are simultaneously subjected to a magnetic field and to an alternating rf electrical field applied perpendicular to the magnetic field, one of which is variable and the other held constant. This combination of fields causes the ions to move in a cycloidal path. When the energy of the variable field reached a value equal to the angular frequency of an ion in the constant field, that ion absorbs energy and increases its angular frequency, moving in a larger cycloidal path. A marginal oscillator detects the increase in energy, and the detector output is plotted against the rate of scan of the variable field to yield a mass spectrum of the sample.[177,178]

A "double resonance" technique is employed to measure the occurrence or nonoccurrence of chemical reaction. Application of a second rf electrical field at the angular frequency of a species suspected of being the reactant source of an observed product ion imparts translational kinetic energy to the reactant. The product ion signal is monitored to observe the effect of heating the suspected reactant. If the product signal increases, the reaction that produces it is endothermic; this implies that the reverse reaction, that is, product to reactant, is exothermic and that the basicity of the deprotonated form of the reactant exceeds that of the deprotonated form of the product. On the other hand, a decrease in product signal upon heating the possible reactant is evidence for exothermic or thermoneutral forward (reactant-to-product) reaction in the absence of heating, that is, the deprotonated form of the product is more basic than the deprotonated form of the reactant. If there is no change in the product signal when the suspected reactant is heated, then there is no chemical reaction linking the two species. Sweeping the second rf electrical field results in a spectrum containing all product ions chemically linked to the reactant ion of interest.[177,178]

ICR data, like MS data, may be either qualitative[137,178–180] or quantitative. Quantitative ICR data is obtained in a manner similar to that in which quantitative MS data is obtained; that is, equilibrium constants are measured, relative PAs are determined, and a proton affinity scale is constructed by stepwise combination of species of known relative proton affinities with species of unknown PA in protolysis equilibria. Entropy effects are neglected. Proton affinities determined by ICR and by high-pressure MS generally agree within several kcal/mole,[181] despite the different experimental temperature and pressure conditions under which measurements are made. The accuracy of relative gas-phase acid–base strengths determined via ICR is equivalent to that obtained from high-pressure MS.[181]

Although enthalpy changes are conventionally regarded as *the* measure of relative gas-phase acidity and basicity, Aue, Webb, and Bowers[141,142,183,184] measured the free-energy analogue of proton affinity, which they defined as the gas-phase basicity GB. GB includes the effect of symmetry number and other changes in entropy on gas-phase acidity and basicity; GB values are thus 8–12 kcal/mole lower than corresponding PA values.

Several applications of gas-phase acid–base measurements merit mention as examples of the way in which revised views of chemical behavior are derived from gas-phase data. The basicities of the alkali metal hydroxides are indistinguishable in solution, but in the gas phase the basicities of this group of compounds increase with increasing electron-releasing ability of the alkali metal cation.[155] The PAs of pyridine and aniline exceed the PA of ammonia, in contrast to the basicity order of these compounds in aqueous solution;[132,156] thus the classical explanations of the "weaker" basicities of

C_5H_5N and $C_6H_5NH_2$ relative to NH_3 in terms of electron withdrawing and resonance effects are erroneous, and the reversals are actually a consequence of preferential NH_3 solvation by water. Solvation reverses the acidity order of simple aliphatic alcohols found in the gas phase[135,136]; the true order is based on charge-delocalizing ability.

Gas-phase acidity–basicity measurements also provide information about substitution effects,[144,151,159,160,181,185,186] data on bond energies, electron and hydrogen affinities, and ionization potentials;[138,162,167,187] and data on entropy effects where significant,[181] for example, in diamines, where cyclization accompanies protonation,[157] or in benzene, where the symmetry number ratio of the conjugates is 12.[164]

One area of extensive application of gas-phase acidity–basicity data is the comparison with solution data for the purpose of quantitating solvation effects. Simple comparison of PA with heat of ionization in solution, or of GB with free energy of ionization in solution, indicates the magnitude of acid–base "strengthening" or "weakening" as a result of solvation.[151,159,161] Characteristic attentuation of acidity or basicity with solvation may sometimes be obtained for a related group of compounds by plotting enthalpies or free energies of ionization in the gas phase against the corresponding quantities in solution. These plots are often linear, and their slopes indicate the magnitude of the "typical" solvation effect for a given class of compounds.[184,185,188–191] For example, a plot of PA versus the heat of protonation in water for amines yields three sets of points; these sets correspond to primary, secondary, and tertiary amines, respectively, and the three "best-fit" lines drawn through them are approximately parallel.[140] However, in cases where reversals in strength occur upon going from the gas phase into solution, no such correlation is possible.[188]

The most thorough investigations of the effects of solvation on acidity and basicity were performed by Arnett and co-workers,[126,184,192–198] who separated the total solvation effect into contributions from each of the members of a conjugate acid–base pair. This approach is based on the thermodynamic cycle

$$
\begin{array}{lccccc}
\text{Gas phase:} & BH^+_{(g)} & \xrightarrow{\Delta P_{i(g)}} & B_{(g)} & + & H^+_{(g)} \\
 & \downarrow \Delta P_S(BH^+) & & \downarrow \Delta P_S(B) & & \downarrow \Delta P_S(H^+) \\
\text{Solution in medium M:} & BM^+_{(M)} & \xrightarrow{\Delta P_{i(M)}} & B_{(M)} & + & H^+_{(M)}
\end{array}
\qquad (107)
$$

where P represents a thermodynamic property (free energy, enthalpy, or entropy), and the subscripts i and S represent ionization and solution, respectively. From equation (107)

$$\Delta P_{i(g)} = \Delta P_{i(M)} - \Delta P_S(\mathrm{B}) - \Delta P_S(\mathrm{H}^+) + \Delta P_S(\mathrm{BH}^+) \qquad (108)$$

It is more convenient from a practical standpoint to work with relative thermodynamic properties $\delta_R \Delta P$ referenced to a standard, for example, ammonia in the case of amines, than with the absolute properties of equation (108), since the $\Delta P_S(\mathrm{H}^+)$ term disappears when relative properties are used. Rearranging equation (108) with relative properties thus yields

$$\delta_R \Delta P_S(\mathrm{BH}^+) = \delta_R \Delta P_{i(g)} + \delta_R \Delta P_S(\mathrm{B}) - \delta_R \Delta P_{i(M)} \qquad (109)$$

The terms on the right side of equation (109) may be determined experimentally. Utilizing gas-phase thermodynamic properties obtained from MS or ICR together with solution thermodynamic properties and $\delta_R \Delta P_S(\mathrm{B})$ determined by calorimetry (see the section on acidity functions for a detailed discussion of Arnett's calorimetric techniques for measuring acid–base strength in solution), Arnett calculated $\delta_R \Delta P_S(\mathrm{BH}^+)$ from equation (109) and was consequently able to explain solvation effects in terms of preferential solvation of either BH^+ or B. An analogous treatment of (0, −1)-charge-type acid–base conjugate pairs yields[199]

$$\delta_R \Delta P_S(\mathrm{B}^-) = \delta_R \Delta P_S(\mathrm{BH}) + \delta_R \Delta P_{i(M)} - \delta_R \Delta P_{i(g)} \qquad (110)$$

Aue, Webb, and Bowers[184] separated $\Delta H_S(\mathrm{BH}^+)$ into size- and charge-dependent contributions; the former is equivalent to the introduction of a neutral species of the same size as BH^+ into solution, and thus cancels approximately with $\Delta H_S(\mathrm{B})$. The interaction of ionic charge with the solvent is thus seen to be significant in the solvation process. When considering free energies of solvation instead of enthalpies, symmetry number and charge-dependent entropy terms may modify predictions based solely on enthalpy effects.

Relative proton affinities of water and alcohols correlate well with the relative oxygen 1*s* core level binding energies determined by X-ray photoemission spectroscopy. The ionization process may be represented as[200]

$$\mathrm{ROH} \rightarrow \mathrm{RO}^{\star}\mathrm{H}^+ + e^- \qquad \Delta H = E_B(1s) \qquad (111)$$

where the ionized electron originates in the 1*s* shell instead of the outer shell, and where $E_B(1s)$ represents the core level binding energy. The analogy between equations (93) and (111) is readily apparent: both illustrate the formation of positively charged ions, one by electron abstraction and the other by proton addition. It is therefore logical to expect similarities between the relative quantitatives ΔPA and $\Delta E_B(1s)$ for a series of compounds. For the ROH series

$$-\Delta E_B(1s) \sim \Delta \mathrm{PA} \tag{112}$$

A similar correlation for nitrogen (1*s*) core level binding energies and amine proton affinities also exists. This relationship was developed into a potential energy model for calculating either ΔPA or $\Delta E_B(1s)$ by Davis et al.[201,202]

Several theoretical models account for substituent effects on proton affinity. A molecular orbital approach[203] separates the substituent effect into electrostatic, exchange (Pauli exclusion), polarization–induction, charge-transfer, and coupling interaction terms; PAs calculated by this method are generally higher than those determined experimentally. Molecular orbital calculations were used to generate substituent effects and theoretical proton affinities for amines,[204–206] substituted benzenes,[206–208] and carboxylic acids.[206] Taft and Levitt[209] correlated gas-phase amine basicity with an empirical set of substituent constants based on polarizability and inductive effects.

2.6 ADVANTAGES AND DISADVANTAGES OF THE BRØNSTED–LOWRY THEORY

The Brønsted–Lowry theory is superior to the Arrhenius concept in many respects. The recognition of acidity and basicity as general phenomena reciprocally related to each other allows extension of acid–base concepts to nonaqueous solutions. One of the most valuable aspects of the protonic theory is its incorporation of useful quantitative classical concepts, with modifications to render them applicable to protolysis in general. Quantitative interpretations of leveling, salt, and medium effects also stem from the protonic concept.

On the other hand, the theory is not sufficiently universal; there are instances of acid–base behavior that are not included, namely, those in which protons and protolysis do not play a part.[9,11,12,15,17] The theory represents almost no advance in this respect over concepts formalized nearly a century prior to its appearance. Aprotic substances manifesting acid behavior, for example, SO_3, CO_2, SiO_2, and BCl_3, are excluded from consideration as acids. Lowry[19] attempted to answer this criticism by describing these substances as acid anhydrides that become Brønsted–Lowry acids upon dissolution in protic solvents. However, Lowry's approach fails to explain the observed acidic properties of these and other substances in aprotic solvents, melts, and in the solid or gaseous state, for example, thionyl chloride in liquid SO_2, $AlCl_3$ in phosgene, and so on. Examples of other ambiguities include the classification of ammonium and mono-, di-, and trialkylammonium ions, but not tetraalkylammonium ions, as acids; also metal ions, for example, Fe^{3+} and Al^{3+}, are not acids, although their complex aquometal cations are. Lowry[19] attempted to integrate the coordination theories of Werner and of Lewis[20] into the protonic theory in order to justify classifying such species as the aquometal ions as acids, seeking to equate the hygroscopicity and consequent increasing acidity of coordination complexes, for example,

$$[Co(NH_3)_3Cl_3] + H_2O \rightleftharpoons [Co(NH_3)_3(H_2O)Cl_2]Cl \qquad (113)$$

with the identical properties of simple acids, utilizing the true acid representation of Hantsch's coupled with Lowry's own intramolecular ionization concept:

$$H\begin{smallmatrix} \cdot\cdot O^- & & O^- \cdot\cdot \\ & S^{2+} & \\ \cdot\cdot O^- & & O^- \cdot\cdot \end{smallmatrix}H + 2H_2O \rightleftharpoons O\begin{smallmatrix} H\cdots O^- & & O^-\cdots H \\ & S^{2+} & \\ H\cdots O^- & & O^-\cdots H \end{smallmatrix}O + 2H^+ \qquad (114)$$

Unfortunately, both the inclusion of aquometal ions, but not bare metal ions, as acids and formulations such as equation (114) are based on the ease of coordination of water and are more difficult to envision in other solvents, reflecting a narrow viewpoint limited to aqueous solution, or precisely what Brønsted and Lowry intended to depart from.

REFERENCES

1. Brønsted, J. N., *Rec. Trav. Chim. Pays-Bas*, **42**, 718 (1923).
2. Brønsted, J. N., *J. Phys. Chem.*, **30**, 777 (1926).
3. Brønsted, J. N., *Ber.*, **61**, 2049 (1928).
4. Brønsted, J. N., *Chem. Rev.*, **5**, 231 (1928).
5. Brønsted, J. N., *Z. Phys. Chem.*, **169**, 52 (1934).
6. Hall, N. F., *J. Chem. Educ.*, **7** 782 (1930).
7. Schwarzenbach, G., *Helv. Chim. Acta*, **13**, 870 (1930).
8. Kolthoff, I. M., *Rec. Trav. Chim. Pays-Bas*, **49**, 401 (1923).
9. Hall, N. F., *J. Chem. Educ.*, **17**, 124 (1940).
10. Luder, W. F., *ibid.*, **25**, 555 (1948).

11. Kolthoff, I. M., *J. Phys. Chem.*, **48**, 51 (1944).
12. Bell, R. P., *Q. Rev. Chem. Soc.*, **1**, 113 (1947).
13. Gillespie, R., "Proton Transfer Reaction," Caldin, E., and Gold, V., Eds., Chapman and Hall, London, 1975, Chap. 1.
14. Dole, M., "Principles of Experimental and Theoretical Electrochemistry," McGraw-Hill, New York, 1925.
15. Day, Jr., M. C., and Selbin, J., "Theoretical Inorganic Chemistry," 2nd ed., Reinhold, New York, 1969.
16. Kolthoff, I. M., "Treatise on Analytical Chemistry," Part I, Vol. I, Kolthoff, I. M., Elving, P. J., and Sandell, E. B., Eds., Wiley-Interscience, New York, 1959, Chap. 11.
17. Gyenes, I., "Titration in Nonaqueous Media," Van Nostrand, Princeton, New Jersey, 1967.
18. Lowry, T. M., *Chem. Ind. (London)*, **42**, 43 (1923).
19. Lowry, T. M., *ibid.*, **42** 1048 (1923).
20. Lewis, G. N., "Valence and the Structure of Atoms and Molecules," Chemical Catalog Co., New York, 1923.
21. Brønsted, J. N., and Wynne-Jones, W. F. K., *Trans. Faraday Soc.*, **25**, 59 (1929).
22. Folin, O., and Flanders, F. F., *J. Am. Chem. Soc.*, **34**, 774 (1912).
23. Kilpatrick, Jr., M., and Rushton, J. H., *J. Phys. Chem.*, **34**, 2180 (1930).
24. Bates, R. G., "Treatise on Analytical Chemistry," Part I, Vol. I, Kolthoff, I. M., Elving, P. J., and Sandell, E. B., Eds., Wiley-Interscience, New York, 1959, Chap. 10.
25. Bruckenstein, S., and Kolthoff, I. M., *ibid.*, Chap. 12.
26. Audrieth, L. F., and Schmidt, M. T., *Proc. Nat. Acad. Sci. U.S.A.*, **20**, 221 (1934).
27. Audrieth, L. F., and Moeller, T., *J. Chem. Educ.*, **20**, 219 (1943).
28. Kolthoff, I. M., and Bruckenstein, S., "Treatise on Analytical Chemistry," Part I, Vol. I, Kolthoff, I. M., Elving, P. J., and Sandell, E. B., Eds., Wiley-Interscience, New York, 1959, Chap. 13.
29. Bates, R. G., "Determination of pH, Theory and Practice," 2nd ed., Wiley, New York, 1973.
30. Bjerrum, N., *Chem. Rev.*, **16**, 287 (1935).
31. Hammett, L. P., *J. Am. Chem. Soc.*, **50**, 2666 (1928).
32. Hall, N. F., and Conant, J. B., *ibid.*, **49**, 3047 (1927).
33. Conant, J. B., and Hall, N. F., *ibid.*, **49**, 3062 (1927).
34. Vander Werf, C. A., "Acids, Bases, and the Chemistry of the Covalent Bond," Reinhold, New York, 1961.
35. Lowry, T. M., *Trans. Faraday Soc.*, **18**, 285 (1923).
36. Lowry, T. M., *ibid.*, **20**, 13 (1924).
37. Lee, T. S., "Treatise on Analytical Chemistry," Part I, Vol. I, Kolthoff, I. M., Elving, P. J., and Sandell, E. B., Eds., Wiley-Interscience, New York, 1959.
38. Hammett, L. P., and Deyrup, A. J., *J. Am. Chem. Soc.*, **54**, 2721 (1932).
39. Bell, R. P., "The Proton in Chemistry," 2nd ed., Cornell University Press, Ithaca, New York, 1973.
40. Popovych, O., "Treatise on Analytical Chemistry," Part I, 2nd ed., Vol. I, Kolthoff, I. M., and Elving, P. J., Eds., Wiley, New York, 1978.

41. Michaelis, L., and Granick, S., *J. Am. Chem. Soc.*, **64**, 1861 (1942).
42. Randles, J. E. B., and Tedder, J. M., *J. Chem. Soc.*, 1218 (1955).
43. Hammett, L. P., *Chem. Rev.*, **16**, 67 (1935).
44. Paul, M. A., and Long, F. A., *ibid.*, **57**, 1 (1957).
45. Jorgensen, J. J., and Hartter, D. R., *ibid.*, **85**, 878 (1963).
46. Paul, M. A., *J. Am. Chem. Soc.*, **72**, 3813 (1950).
47. Yates, K., and Wai, H., *ibid.*, **86**, 5408 (1964).
48. Hammett, L. P., and Paul, M. A., *ibid.*, **56**, 827 (1934).
49. Kilpatrick, M., and Hyman, H. H., *ibid.*, **80**, 77 (1958).
50. Arnett, E. M., and Wu, C. Y., *ibid.*, **82**, 4999 (1960).
51. Arnett, E. M., and Wu, C. Y., *ibid.*, **82**, 5660 (1960).
52. Gold, V., and Hawes, B. W. V., *J. Chem. Soc.*, 2102 (1951).
53. William, G., and Bevan, M., *Chem. Ind.* (*London*), 171 (1955).
54. Gold, V., *ibid.*, 172 (1955).
55. Gold, V., *J. Chem. Soc.*, 1263 (1955).
56. Deno, N. C., Jaruzelski, J. J., and Schriesheim, A., *J. Amer. Chem. Soc.*, **77**, 3044 (1951).
57. Deno, N. C., Berkheimer, H. E., Evans, W. L., and Peterson, H. J., *ibid.*, **81**, 2344 (1959).
58. Murray, M. A., and Williams, G., *J. Chem. Soc.*, 3322 (1950).
59. Deno N. C., Groves, P. T., and Saines, G., *J. Amer. Chem. Soc.*, **81**, 5790 (1959).
60. Boyd, R. H., *ibid.*, **82**, 4288 (1961).
61. Stewart, R., and O'Donnell, J. P., *ibid.*, **84**, 493 (1962).
62. Veleshik, P., Bielavsky, J., and Vecery, M., *Collect. Czech. Chem. Commun.*, **33**, 1687 (1968).
63. Boyd, R. H., *J. Am. Chem. Soc.*, **85**, 1555 (1963).
64. Yates, K., and McClelland, R. A., *ibid.*, **89**, 2686 (1967).
65. Hammett, L. P., and Deyrup, A. J., *ibid.*, **54**, 4239 (1932).
66. Hall, N. H., and Spengemann, W. F., *ibid.*, **62**, 2487 (1940).
67. Spengemann, W. F., and Hall, N. H., *ibid.*, **62**, 2493 (1940).
68. Lemaire, H., and Lucas, H. J., *ibid.*, **73**, 5198 (1951).
69. Bruckenstein, S., *ibid.*, **82**, 307 (1960).
70. Braude, E. A., *J. Chem. Soc.*, 1871 (1948).
71. Braude, E. A., and Stern, E. S., *ibid.*, 1976 (1948).
72. Sanders, W. N., and Berger, J. E., *Anal. Chem.*, **39**, 1473 (1967).
73. Deno, N. C., *J. Am. Chem. Soc.*, **74**, 2039 (1952).
74. Jaffé, H. H., and Gardener, R. W., *ibid.*, **80**, 319 (1958).
75. Yeh, S. J., and Jaffé, H. H., *ibid.*, **81**, 3274 (1959).
76. Kolthoff, I. M., and Bruckenstein, S., *ibid.*, **78**, 1 (1956).
77. Noyce, D. S., and Pryor, W. A., *ibid.*, **77**, 1397 (1955).
78. Rocek, J., *Chem. Listy*, **50**, 726 (1956).
79. Hogfeldt, E., and Bigeleisen, J., *J. Am. Chem. Soc.*, **82**, 15 (1960).
80. Lewis, G. N., and Bigeleisen, J., *ibid.*, **65**, 1144 (1943).
81. Coryell, C. D., and Fix, R. C., *J. Inorg. Nucl. Chem.*, **1**, 119 (1955).

82. Gillespie, R. J., *Chem. Can.*, **19**, 39 (1967).
83. Long, F. A., and Paul, M. A., *Chem. Rev.*, **57**, 935 (1957).
84. Stewart, R., and Mathews, T., *Can. J. Chem.*, **38**, 602 (1960).
85. Johnson, C. D., Katritzky, A. R., and Shapiro, S. A., *J. Am. Chem. Soc.*, **91**, 6654 (1969).
86. Attiga, S. A., and Rochester, C. H., *J. Chem. Soc. Perkin Trans. 2*, 1411 (1975).
87. Tickle, P., Briggs, A. G., and Wilson, J. M., *J. Chem. Soc. B*, 65 (1970).
88. Paul, M. A., *J. Am. Chem. Soc.*, **76**, 3236 (1954).
89. McIntyre, D., and Long, F. A., *ibid.*, **76**, 3240 (1954).
90. Long, F. A., and McIntyre, D., *ibid.*, **76**, 3243 (1954).
91. Rosenthal, D., and Dwyer, J. S., *Can. J. Chem.*, **41**, 80 (1963).
92. Boyer, J. P. H., Corria, R. J. P., Perz, R. J. M., and Reye, C. G., *Tetrahedron*, **37**, 377 (1975).
93. Deno, N. C., and Taft, Jr., R. W., *J. Am. Chem. Soc.*, **76**, 244 (1954).
94. Deno, N. C., and Perizzolo, C., *ibid.*, **79**, 1345 (1957).
95. Taft, Jr., R. W., *ibid.*, **82**, 2965 (1960).
96. Schubert, W. M., Burkett, H., and Schy, A. L. *ibid.*, **86**, 2520 (1964).
97. Wyatt, P. A. H., *Discuss. Faraday Soc.*, **24**, 162 (1957).
98. Bascombe, K. N., and Bell, R. P., *ibid.*, **24**, 158 (1957).
99. Hogfeldt, E., *Acta Chem. Scand.*, **14**, 1627 (1960).
100. Perrin, C., *J. Am. Chem. Soc.*, **86**, 256 (1964).
101. Hammett, L. P., and Paul, M. A., *ibid.*, **56**, 830 (1934).
102. Hammett, L. P., and Zucker, L., *ibid.*, **61**, 2791 (1939).
103. Long, F. A., *Proc. Chem. Soc.*, 220 (1957).
104. Deno, N. C., *J. Am. Chem. Soc.*, **74**, 2039 (1952).
105. Long. F. A., and Pritchard, J. G., *ibid.*, **78**, 2667 (1956).
106. Hammett, L. P., and Paul, M. A., *ibid.*, **58**, 2182 (1936).
107. Chiang, Y., and Whipple, E. B., *ibid.*, **85**, 2763 (1963).
108. Bunnett, J. F., *ibid.*, **83**, 4956 (1961).
109. Bunnett, J. F., *ibid.*, **83**, 4968 (1961).
110. Bunnett, J. F., *ibid.*, **83**, 4973 (1961).
111. Bunnett, J. F., *ibid.*, **83**, 4978 (1961).
112. Deno, N. C., and Stein, R., *ibid.*, **78**, 578 (1956).
113. Lowen, A. M., Murray, M. A., and Williams, G., *J. Chem. Soc.*, 3318 (1950).
114. Hammett, L. P., *Chem. Rev.*, **13**, 61 (1933).
115. Bascombe, K. N., and Bell, R. P., *J. Chem. Soc.*, 1096 (1959).
116. Noyce, D. S., and Jorgensen, M. J., *J. Am. Chem. Soc.*, **84**, 4312 (1962).
117. Gutbezahl, B., and Grunwald, E., *ibid.*, **75**, 559 (1953).
118. Arnett, E. M., and Mach, G. W., *ibid.*, **86**, 2671 (1964).
119. Arnett, E. M., and Mach, G. W., *ibid.*, **88**, 1177 (1966).
120. Arnett, E. M., and Burke, J. J., *ibid.*, **88**, 4308 (1966).
121. Arnett, E. M., Quirk, R. P., and Burke, J. J., *ibid.*, **92**, 1260 (1970).
122. Arnett, E. M., Carter, J. V., and Quirk, R. P., *ibid.*, **92**, 1770 (1970).

123. Arnett, E. M., and Quirk, R. P., *ibid.*, **92**, 3977 (1970).
124. Arnett, E. M., Moriarity, J. C., Small, L. E., and Rudolph, J. P., *ibid.*, **95**, 1492 (1973).
125. Arnett, E. M., Johnston, D. E., Small, L. E., and Oancea, D., *Faraday Symp. Chem. Soc.*, **10**, 20 (1975, published 1976).
126. Arnett, E. M., and Small, L. E., *J. Am. Chem. Soc.*, **99**, 808 (1977).
127. Sherman, J., *Chem. Rev.*, **11**, 93 (1932).
128. Ritchie, C. D., and Uschold, R. W., *J. Am. Chem. Soc.*, **90**, 2821 (1968).
129. Haberfield, P., and Rakshit, A. K., *ibid.*, **98**, 4393 (1976).
130. Munson, M. S. B., *ibid.*, **87**, 2332 (1965).
131. Kebarle, P., and McMahon, T. B., *ibid.*, **98**, 3399 (1976).
132. Yamdagni, R., and Kebarle, P., *ibid.*, **93**, 7139 (1971).
133. Cumming, J. B., and Kebarle, P., *Can. J. Chem.*, **56**, 1 (1978).
134. McMahon, T. B., and Kebarle, P., *J. Am. Chem. Soc.*, **99**, 2222 (1977).
135. Braumann, J. I., and Blair, L. K., *ibid.*, **90**, 6561 (1968).
136. Braumann, J. I., and Blair, L. K., *ibid.*, **92**, 5986 (1970).
137. Braumann, J. I., and Blair, L. K., *ibid.*, **93**, 3911 (1971).
138. Holtz, D., and Beauchamp, J. L., *ibid.*, **91**, 5913 (1969).
139. Braumann, J. I., and Eyler, J. R., Blair, L. K., White, M. J., Comisarow, M. B., and Smyth, K. C., *ibid.*, **93**, 6360 (1971).
140. McIver, Jr., R. T., and Miller, J. S., *ibid.*, **96**, 4323 (1974).
141. Aue, D. H., Webb, H. M., and Bowers, M. T., *ibid.*, **94**, 4726 (1972).
142. Aue, D. H., Webb, H. M., and Bowers, M. T., *ibid.*, **98**, 311 (1976).
143. Staley, R. H., Kleckner, J. E., and Beauchamp, J. L., *ibid.*, **98**, 2081 (1976).
144. Henderson, W. G., Taagepera, M., Holtz, D., McIver, Jr. R. T., Beauchamp, J. L., and Taft, R. W., *ibid.*, **94**, 4728 (1972).
145. Staley, R. H., and Beauchamp, J. L., *ibid.*, **96**, 6252 (1974).
146. Haney, M. A., and Franklin, J. L., *J. Chem. Phys.*, **50**, 2028 (1969).
147. Harrison, A. G., Irko, A., and Van Raalte, D., *Can. J. Chem.*, **44**, 1625 (1966).
148. Haney, M. A., and Franklin, J. L., *J. Phys. Chem.*, **73**, 4328 (1969).
149. Tal'rose, V. L., and Frankevich, E. L., *J. Am. Chem. Soc.*, **80**, 2344 (1958).
150. Tal'rose, V. L., *Pure Appl. Chem.*, **5**, 455 (1962).
151. Yamdagni, R., and Kebarle, P., *J. Am. Chem. Soc.*, **95**, 4050 (1973).
152. Long, J., and Munson, M. S. B., *ibid.*, **95**, 2427 (1973).
153. Beggs, D. P., and Field, F. H., *ibid.*, **93**, 1576 (1971).
154. Field, F. H., and Beggs, D. P., *ibid.*, **93**, 1585 (1971).
155. Searles, S. K., Dzidic, I., and Kebarle, P., *ibid.*, **91**, 2810 (1969).
156. Briggs, J. P., Yamdagni, R., and Kebarle, P., *ibid.*, **94**, 5128 (1972).
157. Yamdagni, R., and Kebarle, P., *ibid.*, **95**, 3504 (1973).
158. Kiraoka, K., Yamdagni, R., and Kebarle, P., *ibid.*, **95**, 6833 (1973).
159. Yamdagni, R., McMahon, T. B., and Kebarle, P., *ibid.*, **96**, 4035 (1974).
160. McMahon, T. B., and Kebarle, P., *ibid.*, **96**, 5940 (1974).
161. Yamdagni, R., and Kebarle, P., *Can. J. Chem.*, **52**, 861 (1974).

162. Cumming, J. B., Magnera, T. F., and Kebarle, P., *ibid*., **55**, 3474 (1972).

163. Kebarle, P., Yamdagni, R., Hiraoka, K., and McMahon, T. B., *Int. J. Mass Spectrom. Ion Phys*., **19**, 71 (1976).

164. Yamdagni, R., and Kebarle, P., *J. Am. Chem Soc*., **98**, 1320 (1976).

165. Kiraoka, K., and Kebarle, P., *ibid*., **98**, 6119 (1976).

166. Lau, Y. K., and Kebarle, P., *ibid*., **98**, 7452 (1976).

167. Cumming, J. B., and Kebarle, P., *ibid*., **99**, 5818 (1977).

168. Kiraoka, K., Grimsrud, E. P., and Kebarle, P., *ibid*., **96**, 3359 (1974).

169. Lau, Y. K., Saluja, P. P. S., Kebarle, P., and Alder, R. W., *ibid*., **100**, 7328 (1978).

170. Bohme, D. K., Fennelly, P., Hemsworth, R. S., and Schiff, H. I., *ibid*., **95**, 7512 (1973).

171. Bohme, D. K., Lee-Ruff, E., and Young, L. B., *ibid*., **93**, 4608 (1971).

172. Bohme, D. K. Lee-Ruff, E., and Young, L. B., *ibid*., **94**, 5153 (1972).

173. Bohme, D. K., and Young, L. B., *ibid*., **92**, 3301 (1970).

174. MacKay, G. I., Betowski, L. D., Payzant, J. D., Schiff, H. I., and Bohme, D. K., *J. Phys. Chem*., **80**, 2919 (1976).

175. Bohme, D. K., Young, L. B., and Lee-Ruff, E., *Can. J. Chem*., **49**, 979 (1971).

176. Dzidic, I., *J. Am. Chem. Soc*., **94**, 8333 (1972).

177. Kriemler, P., and Buttrill, Jr., S. E., *ibid*., **92**, 1123 (1970).

178. Braumann, J. I., and Blair, L. K., *ibid*., **90**, 5636 (1968).

179. Braumann, J. I., and Blair, L. K., *ibid*., **91**, 2126 (1969).

180. Braumann, J. I., Riveros, J. M., and Blair, L. K., *ibid*., **93**, 3914 (1971).

181. Wolf, J. F., Staley, R. H., Taagepera, M., McIver, Jr., R. T., Beauchamp, J. L., and Taft, R. W., *ibid*., **99**, 5417 (1977).

182. Bowers, M. T., Aue, D. H., Webb, H. M., and McIver, Jr., R. T., *ibid*., **93**, 4314 (1971).

183. Aue, D. H., Webb, H. M., and Bowers, M. T., *ibid*., **95**, 2699 (1973).

184. Aue, D. H., Webb, H. M., and Bowers, M. T., *ibid*., **98**, 318 (1976).

185. Taagepera, M., Henderson, W. G., Brownlee, R. T. C., Beauchamp, J. L., Holtz, D., and Taft, R. W., *ibid*., **94**, 1369 (1972).

186. Staley, R. H., Taagepera, M., Henderson, W. G., Koppel, I., Beauchamp, J. L., and Taft, R. W., *ibid*., **99**, 326 (1977).

187. Beauchamp, J. L., Holtz, D., Woodgate, S. D., and Patt, S. L., *ibid*., **94**, 2798 (1972).

188. McIver, Jr., R. T., and Silvers, J. H. *ibid*., **95**, 8462 (1973).

189. Bordwell, F. G., Bartmess, J. E., Drucker, G. E., Margolin, Z., and Matthews, W. S., *ibid*., **97**, 3226 (1975).

190. Aue, D. H., Webb, H. M., Bowers, M. T., Liotta, C. L., Alexander, C. J., and Hopkins, Jr., H. P., *ibid*., **98**, 854 (1976).

191. Wolf, J. F., Harch, P. G., and Taft, R. W., *ibid*., **97**, 2904 (1975).

192. Arnett, E. M., Jones III, F. M., Taagepera, M., Henderson, W. G., Beauchamp, J. L., Holtz, D., and Taft, R. W., *ibid*., **94**, 4724 (1972).

193. Arnett, E. M., and Wolf, J. F., *ibid*., **97**, 3262 (1975).

194. Arnett, E. M., and Abboud, J. M., *ibid*., **97**, 3865 (1975).

195. Arnett, E. M., Johnston, D. E., and Small, L. E., *ibid*., **97**, 5598 (1975).

196. Arnett, E. M., Small, L. E., Oancea, D., and Johnston, D., *ibid*., **98**, 7346 (1976).

197. Arnett, E. M., Chawla, B., Bell, L., Taagepera, M., Hehre, W. J., and Taft, R. W., *ibid.*, **99**, 5729 (1977).

198. Arnett, E. M., *Acc. Chem. Res.*, **6**, 404 (1973).

199. Arnett, E. M., Small, L. E., McIver, Jr., R. T., and Miller, J. S., *J. Am. Chem. Soc.*, **96**, 5638 (1974).

200. Martin, R. L., and Shirley, D. A., *ibid.*, **96**, 5299 (1974).

201. Davis, D. W., and Rabalais, J. W., *ibid.*, **96**, 5305 (1974).

202. Davis, D. W., and Shirley, D. A., *ibid.*, **98**, 7898 (1976).

203. Umeyana, H., and Morokuma, K., *ibid.*, **98**, 4400 (1976).

204. Kollman, P., and Rothenberg, S., *ibid.*, **99**, 1333 (1977).

205. Taagepera, M., Hehre, W. J., Topsom, R. D., and Taft, R. W., *ibid.*, **98**, 7438 (1976).

206. Reynold, W. F., Mezey, P. G., Hehre, W. J., Topsom, R. D., and Taft, R. W., *ibid.*, **99**, 5821 (1977).

207. Hehre, W. J., and Pople, J. A., *ibid.*, **94**, 6901 (1972).

208. McKelvey, J. M., Alexandratos, S., Streitweiser, Jr., A., Abboud, J. M., and Hehre, W. J., *ibid.*, **98**, 244 (1976).

209. Taft, R. W., and Levitt, L. S., *J. Org. Chem.*, **42**, 916 (1977).

CHAPTER

3

SOLVENT SYSTEMS THEORY

A major weakness of the Arrhenius theory was the restriction of acid–base concepts to aqueous solution; that is, in spite of experimental evidence that acid–base interactions occurred in other solvents and even in the absence of a solvent, the characteristic classical neutralization process was

$$\text{Acid} + \text{Base} \rightarrow \text{Salt} + \text{Water} \tag{1}$$

The Brønsted–Lowry theory expanded acid–base reactions to include other solvents by introducing a general protolysis scheme in which neutralization was one type of protolytic reaction. Another consequence of protolysis was the replacement of salt formation by conjugate acid and base formation in neutralization reactions.

However, even before the Brønsted and Lowry concepts were published, another extension of acid–base concepts to nonaqueous solvents was being implemented. This approach was predicated on the retention of salt formation as the definitive acid–base process in any solvent by generalizing equation (1):

$$\text{Acid} + \text{Base} \rightarrow \text{Salt} + \text{Solvent} \tag{2}$$

This concept continued to develop during and after the emergence of the protonic theory. Equation (2) implies a different Arrhenius-type acid–base system for each solvent,[1] and this concept is therefore referred to as the theory of solvent systems.

3.1 REACTIONS IN LIQUID AMMONIA AND OTHER PROTOLYTIC SOLVENTS

The physical and chemical properties of liquid ammonia resemble those of water; ammonia has a relatively high boiling point, specific heat, heat of fusion, heat of vaporization, dielectric constant, solvating power, and ionizing power, as does water. The chemical behavior of solutions of ammonium ions in ammonia towards alkali and alkaline-earth metals and towards solvent-insoluble oxides and hydroxides parallels that of aqueous acid (hydronium ion) solutions[2–5]:

$$Mg + 2H_3O^+ \rightarrow Mg^{2+} + H_2 + 2H_2O \tag{3}$$

$$Mg + 2NH_4^+ \rightarrow Mg^{2+} + H_2 + 2NH_3 \tag{4}$$

$$CaO + 2H_3O^+ \rightarrow Ca^{2+} + 3H_2O \tag{5}$$

$$CaO + 2NH_4^+ \rightarrow Ca^{2+} + H_2O + 2NH_3 \tag{6}$$

Alkaline solutions in ammonia impart a red color to phenolpthalein that disappears upon the addition of acid.*

Franklin[2,3] regarded water and ammonia as the parent compounds of analogous families of substances and devised a system of acids, bases, and salts in ammonia paralleling the Arrhenius theory in water.

Ammonia, like water, is capable of autoionization in the pure liquid state[5,6]:

$$2H_2O \rightleftharpoons H_3O^+ + OH^- \tag{7}$$

$$2NH_3 \rightleftharpoons NH_4^+ + NH_2^- \tag{8}$$

The lyonium ions of both solvents are similar in that both are actually solvated protons.[1] Acids may therefore be defined in liquid ammonia by analogy with the Arrhenius theory; that is, an acid is a hydrogen-containing substance that splits off protons in ammonia:

$$HA + H_2O \rightleftharpoons H_3O^+ + A^- \tag{9}$$

$$HA + NH_3 \rightleftharpoons NH_4^+ + A^- \tag{10}$$

*Pure ammonia does not impart a red color to the indicator, as does an ammoniacal, aqueous solution; NH_3 is a base relative to water, but is "neutral" insofar as processes that restrict its role to a solvent are concerned.

Many of the concepts applicable to aqueous solutions of acids may be adapted to solutions of acids in ammonia; for example, any acid stronger than the ammonium ion is leveled to NH_4^+. Acids that are weak in water often behave like strong acids in ammonia, because the latter is a more basic solvent; for example, acetic acid is completely dissociated in ammonia.

Franklin distinguished between two categories of acids for the purpose of making comparisons between solvents on a theoretical basis. On the one hand, there are the "solvoacids," which are solvent derivatives, and on the other hand, there are acids that are not solvent derivatives.

Ammonoacids[2,3] include the ammono analogues of aquo-organic acids and of some nonmetal aquo-inorganic acids, as well as their partially deammonated products. The ammono analogue of acetic acid, $CH_3C(NH_2)$ $=NH$, and the anammonide corresponding to acetic anhydride, $(CH_3C=$ $NH)_2NH$, are representative of organic ammonoacids and related compounds. Inorganic ammonoacids and anammonides are derived from ortho ammonoacids, that is, theoretical combinations of a nonmetallic atom with a number of solvent anions equal to the valence of the nonmetallic atom; desolvation of the ortho ammonoacids yields the characteristic ammonoacids and anammonides. Inorganic aquoacids and acid anhydrides are similarly derived, as shown below for tetravalent carbon, pentavalent nitrogen, hexavalent sulfur, and heptavalent chlorine:

$$C(OH)_4 \xrightarrow{-H_2O} H_2CO_3 \xrightarrow{-H_2O} CO_2 \quad (11)$$

$$N(OH)_5 \xrightarrow{-2H_2O} NHO_3 \xrightarrow{+HNO_3,\,-H_2O} N_2O_5 \quad (12)$$

$$S(OH)_6 \xrightarrow{-2H_2O} H_2SO_4 \xrightarrow{-H_2O} SO_3$$

$$H_2SO_4 \xrightarrow{+H_2SO_4,\,-H_2O} H_2S_2O_7 \quad (13)$$

$$Cl(OH)_7 \xrightarrow{-3H_2O} HClO_4 \xrightarrow{+HClO_4,\,-H_2O} Cl_2O_7 \quad (14)$$

Analogously, for carbon and nitrogen in ammonia,

$$C(NH_2)_4 \xrightarrow{-2NH_3} NCNH_2 \xrightarrow{+NCNH_2,\,-NH_3} (NC)_2NH \xrightarrow{+2(NC)_2NH,\,-NH_3} 2C_3N_4 \quad (15)$$

$$N(NH_2)_5 \xrightarrow{-3NH_3} HN_3 \quad (16)$$

Therefore the ammono analogue of carbon dioxide is C_3N_4, and that of nitric acid is hydrazoic acid (HN_3).

Ammonoacids are generally too weak to manifest acidic tendencies in aqueous solution; in fact, they are often basic in aqueous solution. Their acidity is enhanced and becomes apparent only in a more basic solvent such as ammonia.

The nonammonoacids behave no differently in ammonia from the ammonoacids. They may also be either organic or inorganic. Organic aquoacids that are weak in aqueous solution behave like strong acids; that is, they are completely dissociated in liquid ammonia, for example, acetic acid. Hydrogen and ammonium halides (which are actually identical in ammonia) also are not ammonoacids; it may be noted that they are also not aquoacids. The ammono analogue of the aquoacid nitric acid [equation (12)] is not ammonium nitrate but, rather, hydrazoic acid [equation (16)], which may also be referred to as ammononitric acid. The relationship between HNO_3 and HN_3 is more than mere formula juggling. Hydrazoic acid performs all of the functions in ammonia that nitric acid is capable of in water: it oxidizes ferrous ion to ferric ion; in combination with HCl it dissolves gold; and its salt KN_3 is a nitridizing agent, just as KNO_3 is an oxidizing agent. It is probable that this relationship would have been overlooked in a less rigorous approach that did not distinguish between ammono and nonammonoacids.

Acidic properties are also exhibited by nonmetallic halides undergoing ammonolysis comparable to aqueous hydrolysis, but these substances are not acids in the Franklin sense:

$$AsCl_3 + 6H_2O \rightarrow As(OH)_3 + 3HCl \tag{17}$$

$$AsCl_3 + 6NH_3 \rightarrow As(NH_2)_3 + 3NH_4Cl \tag{18}$$

Ammonobases include metallic amides, imides, and nitrides, for example, KNH_2, K_2NH, and K_3N, that are analogues of the aquobases hydroxide and oxide, for example, KOH and K_2O. Ammonobases possess indicator, solubility, and solid-state conduction properties similar to those of their aquo analogues.[2,3,5,7] Bases stronger than the amide ion are leveled to NH_2^-. Bases that are relatively strong in aqueous solution are often relatively weak in the more basic solvent.

Neutralization processes in liquid ammonia also parallel those in aqueous solution:

$$H_3OCl + NaOH \rightarrow NaCl + H_2O \tag{19}$$

$$NH_4Cl + NaNH_2 \rightarrow NaCl + 2NH_3 \tag{20}$$

$$CH_3COOH + NaOH \rightarrow CH_3COONa + H_2O \quad (21)$$

$$CH_3C(NH_2){=}NH + NaNH_2 \rightarrow CH_3C(NHNa){=}NH + NH_3 \quad (22)$$

Ammonoacids and ammonobases [equation (22)] can react to form ammonosalts, although not all acid–base reactions in liquid ammonia result in ammonosalt formation [equation (20)].

Amphoterism is observed in the amides of elements whose hydroxides are amphoteric in water[3,8]:

$$Zn(NH_2)_2 + 2KNH_2 \rightarrow K_2Zn(NH_2)_4 \quad (23)$$

$$Zn(NH_2)_2 + 2NH_4Cl \rightarrow ZnCl_2 + 2NH_3 \quad (24)$$

These concepts can be extended, at least in principle, to other protic solvents, for example, H_2S, HF, CH_3COOH, H_2SO_4, and even CH_4.[3,5,6] Experimental considerations, such as solubility and hydrolysis by traces of water, often make it difficult to verify Franklin's approach in general. The acidity and basicity of solvents may be compared by considering the behavior of a single solute in a variety of media; for example, urea is acidic in ammonia, weakly basic in water, and is a strong base in glacial acetic acid.[6]

Among the criticisms of Franklin's approach is its restriction to protic solvents. Although more general than the Arrhenius theory, the Franklin concept does not account for aprotic acid–base behavior and in this respect offers no advantage over the Brønsted–Lowry theory. On the contrary, it is simpler to treat protonic acid–base behavior with the single acid–base definition of the latter concept than to define a separate acid–base system for each solvent.

In addition, Franklin's distinction between solvo and nonsolvoacids implies that tautomeric substances are simultaneously members of two solvoacid classes; he regarded acetamide as a mixed aquoammonoacid:[3]

$$CH_3C\begin{matrix}{\nearrow}O \\ {\searrow}NH_2\end{matrix} \rightleftharpoons CH_3C\begin{matrix}{\nearrow}OH \\ {\searrow}NH\end{matrix} \quad (25)$$

and keto-enol tautomers and nitro-organics as mixed aquomethanoacids:

$$CH_3C\begin{matrix}{\nearrow}O \\ {\searrow}CH_3\end{matrix} \rightleftharpoons CH_3C\begin{matrix}{\nearrow}OH \\ {\searrow}CH_2\end{matrix} \quad (26)$$

$$CH_3NO_2 \rightleftharpoons CH_2NOOH \quad (27)$$

The last example coincidentally served as Hantsch's starting point in the development of pseudoacid theory, which began at about the same time that Franklin initially formulated the solvent systems theory.

3.2 APROTIC ACID–BASE REACTIONS

Germann[9,10] agreed with Franklin that hydroxide ion was not the only basic species, but he felt that the restriction of acidic species to hydrogen-containing substances and the consequent view of acid–base processes as proton–anion reactions was only a partial improvement over the Arrhenius theory. Germann questioned the belief that hydrogen was the only carrier of acidic properties, whether in a molecule or as a free or solvated ion, and pointed out similarities between the behavior of sulfur trioxide in water and aluminum chloride in phosgene ($COCl_2$) to illustrate his point (Table 3.1). He concluded that phosgene, like water, ammonia, and other protic solvents, is an ionizing solvent despite the fact that it does not contain hydrogen, although he did not specify the manner of ionization[6,7]:

$$COCl_2 \rightleftharpoons CO^{2+} + 2Cl^- \tag{28}$$

$$COCl_2 \rightleftharpoons COCl^+ + Cl^- \tag{29}$$

$$2COCl_2 \rightleftharpoons COCl^+ + COCl_3^- \tag{30}$$

According to Germann, the similarities in chemical behavior between ionizing solvents indicate that any ionizing solvent can serve as the "parent solvent" of an acid–base–salt system, leading to a generalized set of acid, base, and salt definitions.[9,10]

A solvoacid is defined as a substance that combines with one or more molecules of the parent solvent to produce an addition compound with the following properties:

1. It is an electrolyte capable of dissociating into a cation identical to the solvent cation and a complex anion composed of atoms from both the solvoacid and the parent solvent.
2. Metals and metal compounds react with the addition compound to produce salts; with metals, a gas characteristic of the solvent cation is displaced.
3. The solvoacid is neutralized by a solvobase, producing a salt and pure solvent.

TABLE 3.1 Comparison of Properties of SO_3 in H_2O and $AlCl_3$ in $COCl_2$

Property	SO_3 in H_2O	$AlCl_3$ in $COCl_2$
Solvation	$SO_3 + H_2O \rightarrow H_2SO_4$	$2AlCl_3 + COCl_2 \rightarrow COAl_2Cl_8$
Electrolytic conductivity	$H_2SO_4 \rightarrow 2H^+ + SO_4{}^{2-}$	$COAl_2Cl_8 \rightarrow CO^{2+} + Al_2Cl_8{}^{2-}$
Electrolysis products	$H_2 + \frac{1}{2}O_2$	$CO + Cl_2$
Reaction with metals	$H_2SO_4 + Mg_{(s)} \rightarrow MgSO_4 + H_{2(g)}$	$COAl_2Cl_8 + Mg_{(s)} \rightarrow MgAl_2Cl_8 + CO_{(g)}$
Reaction with metal compounds	$H_2SO_4 + Na_2CO_3 \rightarrow Na_2SO_4 + H_2O + CO_{2(g)}$	$COAl_2Cl_8 + Na_2CO_3 \rightarrow Na_2Al_2Cl_8 + 2CO_{2(g)}$
Acid–base neutralization	$H_2SO_4 + Na_2O \rightarrow \underset{\text{(Salt)}}{Na_2SO_4} + \underset{\text{(Solvent)}}{H_2O}$	$COAl_2Cl_8 + 2NaCl \rightarrow \underset{\text{(Salt)}}{Na_2Al_2Cl_8} + \underset{\text{(Solvent)}}{COCl_2}$

The reactions of aqueous SO_3 and $AlCl_3$ in phosgene exemplify the characteristic behavior of acids according to Germann.

Solvobases are metallic derivatives of the parent solvent that dissociate into metal ions and anions identical to the solvent anion. Solvobases are good conductors and neutralize solvoacids. Solvosalts are the products of solvoacid–metal, solvoacid–metal compound, or solvoacid–solvobase reactions. They are good conductors and may undergo solvolysis.

Germann's definitions rest on the formation of characteristic solvent ions. Substances that increase the CO^{2+} concentration in phosgene are acids, and those that increase the Cl^- concentration are bases. These definitions retain Franklin's solvoacid–nonsolvoacid distinction in name, but alter its meaning to reflect Germann's emphasis on solvation prior to ionization. Hydrogen chloride is a solvoacid in water by this definition, since the hydronium ion concentration of an aqueous HCl solution exceeds that of pure water, whereas Franklin did not regard HCl as an aquoacid. However, HCl is not a solvoacid in phosgene; in fact, one of the ambiguities of Germann's definitions is that HCl in phosgene may be regarded as a base, since it contains the chloride ion. On the other hand, Germann's early solvoacid definition does offer an explanation of metal ion, for example, Al^{3+}, acidity, based on solvation, that is analogous to the Brønsted–Lowry conception of aquometal ion acidity.

Tracer studies with radioactive chlorine[11-13] indicate that the exchange rate between $AlCl_3$ and phosgene in solution is very slow. Heterogeneous chlorine exchange between the pure solvent and ionic chlorides, which are insoluble in phosgene, is also extremely slow. Ionic chlorides exchange rapidly with a phosgene solution of aluminum chloride, but this exchange seems to occur directly between the solute and the solid phase, since no activity is found in solvent distilled from such solutions. This evidence shows that phosgene is not an ionizing solvent, since exchange of chloride with $AlCl_3$ and ionic chlorides would otherwise be rapid. Huston and Lang[12] attributed the conductivities measured by Germann to reactions in which the solvent plays no role, for example,

$$2AlCl_3 \rightleftharpoons AlCl_2^+ + AlCl_4^- \tag{31}$$

However, Germann's unfortunate choice of solvent was not discovered until much later and consequently did not affect other attempted solvent system definitions based on his work.

Cady and Elsey[1,5,14,15] reached similar definitions, describing an acid as a solute that, by direct dissociation in or reaction with an ionizing solvent, increases the solvent cation concentration. A base increases the solvent anion concentration in an analogous manner. Neutralization [equation (2)] was amended by the Cady–Elsey definitions to emphasize the solvent role:

$$\text{Acidic solution} + \text{Basic solution} \rightarrow \text{Salt} + \text{Solvent} \qquad (32)$$

Thus acids do not react with bases directly, but, instead, solvent cations and anions are the actual reacting entities in neutralization. Cady and Elsey defined salts as electrolytes dissociating into at least one cation and at least one anion different from the solvent ions. Acidic and basic salts dissociate to form the solvent cation and anion in addition to the nonsolvent cation and anion, respectively. Other salts are considered to be neutral, although solvolysis of such salts may result in solutions that are not neutral.

Theoretically any solvent may be treated in the same manner as water by this approach, starting with the determination of an autoionization constant K_S and the establishment of a scale analogous to the pH scale, with a point of neutrality at $\frac{1}{2}\text{p}K_S$.

Cady and Elsey retained Franklin's concept of solvoacids and solvobases in their definitions, but the complexities involved in the determination of solvoacids and solvobases in the wide variety of solvents covered by their concept and the functional uselessness of distinguishing between substances with identical behavior relegated this aspect of the solvent systems theory to a position of marginal significance. By the 1930s, when Jander[5,6,8,16–18] applied the Cady–Elsey solvent systems concept to liquid sulfur dioxide, the idea of maintaining such a distinction was obsolete.

The low conductivity of pure liquid SO_2, its good solvating power, and the high conductivity of its solutions caused Jander to view SO_2 as an ionizing solvent whose exact manner of ionization is unknown:

$$SO_2 \rightleftharpoons SO^{2+} + O^{2-} \qquad (33)$$

$$2SO_2 \rightleftharpoons SO^{2+} + SO_3^{\,2-} \qquad (34)$$

$$3SO_2 \rightleftharpoons SOSO_2^{\,2+} + SO_3^{\,2-} \qquad (35)$$

However, his observation that thionyl compounds behave like acids, and sulfites like bases, in this solvent led him to conclude that the solvent cation is the free or solvated thionyl (SO^{2+}) ion, and that the solvent anion is the sulfite ion. Thus thionyl chloride is an acid in liquid SO_2,

$$SOCl_2 \rightleftharpoons SO^{2+} + 2Cl^- \qquad (36)$$

and neutralization in this solvent is represented, in Jander's view, by, for example,

$$SOCl_2 + Cs_2SO_3 \rightarrow 2CsCl + 2SO_2 \tag{37}$$

Substances other than sulfites that are capable of interacting with the solvent to increase the sulfite ion concentration are also considered to be bases,[8] for example,

$$2NH_3 + 2SO_2 \rightleftharpoons (NH_3)_2SO^{2+} + SO_3^{2-} \tag{38}$$

Other characteristic acid–base phenomena are also observed in this solvent; for example, $Al_2(SO_4)_3$ is amphoteric in SO_2, just as aluminum hydroxide is in water.

Exchange experiments with labeled sulfur and oxygen in this medium have disputed the validity of the Cady–Elsey and Jander concepts.[5,19–22] Thionyl chloride and bromide exchange oxygen and sulfur with the solvent very slowly, in contradiction with the assumptions of equations (33)–(35) and (38). On the other hand, sulfite exchange is so rapid in this solvent that it seems unlikely that the limited solvent dissociation alone can account for it. The identity of the solvent cation in SO_2 is therefore uncertain, although the solvent anion is sulfite.

The failure of the Germann and Cady–Elsey concepts in phosgene, sulfur dioxide, and in other solvents raised the question as to whether acids and bases were to be regarded as substances whose characteristic properties are linked to the existence of ions of a solvent without seeking a trait common to and inherent in the acids and bases themselves. Lewis[23] touched on the idea of considering an acid as a substance leading to solvent cation production, or as a substance combining with solvent anions, and regarding a base as capable of solvent anion production or combination with solvent cations. He credited this approach with being more general than the proton definition, but at the same time noted the attempted restriction of what he felt were solvent-independent phenomena. On the other hand, there were extreme adherents of the solvent systems theory as presented by Cady and Elsey who, despite evidence to the contrary, insisted that acidity and basicity were solution properties. They rejected the idea of acid–base reactions in the absence of or without the participation of a solvent, attributing such processes to ionic or polar covalent forces, and maintained that the only true acid–base reactions were those between solvent cations and solvent anions.[24] According to this viewpoint, aqueous hydrochloric and sulfuric acids are acids, but pure hydrogen chloride and hydrogen sulfate are not; this represents a regression to the days of Arrhenius and is recognized as such today.

3.3 ELECTRONIC STRUCTURE AND THE SOLVENT SYSTEMS THEORY

Wickert[8,17,25] formulated an acid–base theory coupling neutralization [equation (2)], which he considered the definitive acid–base property, with a classification of acids and bases founded upon the electronic structures of their ions. Criticizing the Brønsted–Lowry approach as invalid in other than protic solvents, Wickert conceived of his theory as applicable to most ionizing solvents, which he termed "water-like." Water-like solvents dissociate into cations without and anions with closed electronic configurations, for example,

$$H_2O \rightleftharpoons H^+ + OH^- \tag{39}$$

$$SbCl_3 \rightleftharpoons Sb^{3+} + 3Cl^- \tag{40}$$

[See also equations (33)–(35).] Wickert's definition of water-like solvents excludes molten salts and bases because their cations, as well as their anions, have closed configurations, for example,

$$NaCl_{(l)} \rightleftharpoons Na^+ + Cl^- \tag{41}$$

$$NaOH_{(l)} \rightleftharpoons Na^+ + OH^- \tag{42}$$

Thus Wickert differentiated between water-like solvents on one hand and salts and bases on the other solely on the basis of the electronic configurations of their cations.

In this conception the difference between salts and bases is reduced to a comparison of solute and solvent anions. A solute that is a salt in most solvents is a base in those solvents that dissociate to form an anion identical to the solute anion, for example, sodium hydroxide is a base in aqueous solution, but a salt in liquid sulfur dioxide. Wickert also noted that acid dissociation results in the formation of open-shell cations and closed-shell anions, just as in the dissociation of water-like solvents. From these considerations emerged an acid–base theory that pictures salts and bases as one pair of closely related categories, and acids and water-like solvents as another. The nature of the solvent anion distinguishes a base from a salt; an acid differs from a water-like solvent simply because the former is a solute and the latter a solvent; that is, all acids in the pure state are water-like solvents.

Consequently, Wickert's view of neutralization appears to be an analogue of Brønsted's double conjugate pair acid–base equilibrium predicated upon electronic configuration:

	Acid +	Base →	Salt	+ Solvent (43)
Closed electronic configuration:	Anion only	Cation and anion	Cation and anion	Anion only

The interchangeability of acids and solvents, or of bases and salts, implies no distinction on the part of this theory between neutralization and solvolysis, just as in the protolysis concept. Wickert considered his definitions as absolute, solvent independent, and extendible to processes outside the liquid state. One of the corollaries derived from Wickert's definitions is the stipulation that any substance that is an acid in one water-like solvent and dissociates in the same manner in another water-like solvent is also an acid in the latter solvent. This principle clarifies the confusion caused by the Germann and Cady–Elsey concepts in which, for instance, HCl is considered a base in phosgene because it dissociates to form the solvent anion. The electronic structure of the dissociation products of HCl make it unequivocally an acid in the Wickert sense. However, the Wickert definitions cause a reaction such as

$$SbBr_3 + 3HCl \rightarrow 3HBr + SbCl_3 \tag{44}$$

to be regarded as a metathetical acid–acid process rather than an acid–base one in $SbCl_3$ solvent.

Wickert's theory, although formulated on a solvent systems basis, does not limit acids and bases to substances influencing solvent ion concentrations. The stipulation that an acid remains an acid no matter what the solvent (assuming that the manner in which the acid dissociates is unchanged) is an affirmation of the general nature of acidity phenomena and of the inherence of acid properties in the acids themselves.

On the other hand, Wickert implies that ionization is a characteristic property of acids and bases and his concept relegates amphoterism to nothing more than differences in the manner of ionization. In addition, a number of substances with acidic properties fall into the category of Wickert bases, because their cations have open electronic configurations, for example, $AlBr_3$, indicating that these definitions are not general enough.[8]

Cruse[18] pointed out that Wickert's ideas also lead to neutralization processes in which the pure solvent is not a product and that the Wickert theory is therefore not a true solvent systems concept; for example, in liquid SO_2

$$2HCl + K_2SO_3 \rightarrow 2KCl + H_2SO_3 \tag{45}$$

Cruse believed that Wickert erred in regarding neutralization as the sole definitive acid–base property from a solvent systems point of view [equation (2)], because acids and bases are capable of reaction in inert solvents such as benzene. However, indicator color changes may be observed for acid–base reactions in inert as well as in ionizing media, and Cruse used indicators to demonstrate what he believed was Brønsted-type behavior in liquid sulfur dioxide. When aniline is dissolved in SO_2, a yellow color appears that is attributed to the formation of a solute-solvent adduct:

$$C_6H_5NH_2 + SO_2 \rightarrow C_6H_5NH_2SO_2 \tag{46}$$

Cruse postulated that the adduct forms as a result of electron exchange between aniline and SO_2 and that acids bind bases via an electron exchange mechanism. The decolorization of an aniline solution in SO_2 by a hydrogen acid,

$$C_6H_5NH_2SO_2 + HA \rightarrow C_6H_5NH_3^+ + A^- + SO_2 \tag{47}$$

represents a double acid–base equilibrium in which the electron exchange mechanism of aniline is transferred from SO_2 to H^+. This represented Brønsted-type behavior to Cruse and confirmed what he perceived as the general validity of the Brønsted–Lowry theory, even in a solvent such as sulfur dioxide, which has neither proton-donating nor proton-accepting properties.

Smith[8,26,27] also devised a set of solvent systems acid–base definitions founded on electronic structure. An acid is a neutral or charged solute acting as an electron pair acceptor towards a molecule or ion of the solvent, and a base is a neutral or charged solute acting as an electron pair donor towards a solvent molecule or ion. Smith's definitions give the same results as the Brønsted–Lowry theory for protic solvents,

$$ClH + {:}OH_2 \rightleftharpoons Cl^- + H{:}OH_2^+ \tag{48}$$

$$H_2N{:}^- + HOH \rightleftharpoons H_2N{:}H + OH^- \tag{49}$$

but they are equally applicable to aprotic solvents such as selenium oxychloride, regarded by Smith as the parent solvent of a solvosystem of acids, bases, and salts[26,27]:

$$SeOCl_2 \rightleftharpoons SeOCl^+ + Cl^- \tag{50}$$

Acids (electron pair acceptors) increase the solvent cation concentration; for example,

$$SO_3 + SeOCl_2 \rightleftharpoons SO_3Cl^- + SeOCl^+ \quad (51)$$

$$SnCl_4 + 2SeOCl_2 \rightleftharpoons SnCl_6^{2-} + 2SeOCl^+ \quad (52)$$

Bases (electron pair donors) increase the solvent anion concentration; Smith's definitions therefore include those of Germann and of Cady and Elsey. Smith drew a distinction between incidental and other bases. The former are bases containing electron-pair-donating groups identical to the solvent anion, for example, metallic chlorides in $SeOCl_2$. Smith found that these bases are weak compared to those in the latter category, for example,

$$C_5H_5N + SeOCl_2 \rightleftharpoons C_5H_5NSeOCl^+ + Cl^- \quad (53)$$

Neutralization results in salt and solvent formation [equation (2)]:

$$SeOCl^+ + SO_3Cl^- + C_5H_5NSeOCl^+ + Cl^- \rightarrow C_5H_5NSO_3 + 2SeOCl_2 \quad (54)$$

The combination of electronic structure with the solvent systems theory represented an attempt to describe acids and bases in functional rather than in constitutive (solvent cation–anion or protonic) terms. It found acceptance, not within the solvent systems theory; but as a corollary of the electronic acid–base theory of Lewis, which does not distinguish between solute and solvent. Most adherents of the solvent systems theory did not regard electronic structure as significant in explaining acid–base behavior and still maintained that solvent ion formation, by whatever means, was the definitive acid–base trait.

3.4 ACIDS AND BASES IN MELTS

Lux[28] believed that melts could act as solvents, albeit in a high-temperature range, and began the development of an acid–base theory for oxide melts that was analogous to the protonic theory. According to Lux, the oxide ion serves the same purpose in oxide melts as does the proton in aqueous solution; that is, its concentration is characteristic of the state of the melt. The negative charge of the oxide ion requires a reversal of donor–acceptor abilities of acids and bases relative to the proton; that is, a base is an oxide donor and an acid is an oxide acceptor[5,28–31]:

$$\text{Base} \rightleftharpoons \text{Acid} + O^{2-} \tag{55}$$

Lux therefore postulated the same kind of inverse acid–base relationship for oxide melts as did Brønsted for protolytes and was able to utilize some of the ideas associated with the protonic theory. For example, sodium oxide is a base whose conjugate acid is the sodium ion, and sulfur trioxide is an acid whose conjugate base is the sulfate ion:

$$Na_2O \rightleftharpoons 2Na^+ + O^{2-} \tag{56}$$

$$SO_3 + O^{2-} \rightleftharpoons SO_4^{2-} \tag{57}$$

An acid–base reaction is an oxide ion transfer between two conjugate acid–base pairs analogous to protolysis:

$$Na_2O + SO_3 \rightleftharpoons 2Na^+ + SO_4^{2-} \tag{58}$$

Melt acidity and basicity depend on the position of equilibria such as equation (58). A sodium sulfate melt is considered to be in the neutral region of oxide concentration, since very little decomposition to SO_3 and Na_2O occurs as a result of the strong acidity of the former (or the weak basicity of its conjugate base). A sodium carbonate melt is slightly alkaline, since carbonate is a stronger base than sulfate, and the decomposition equilibrium is slightly shifted towards CO_2 and Na_2O formation. A sodium oxide melt is very basic (since sodium ion is an extremely weak conjugate acid).

Lux defined a concentration scale for oxide ion in melts. Having no information about activity coefficients under these circumstances, he set both the activity and concentration of oxide ion equal to the analytical concentration of dissolved oxide, which in melts of known composition is the mole fraction of oxide,

$$n = \frac{n_{Na_2O}}{\sum_i n_i} \tag{59}$$

and, in melts of unknown composition the number of moles of oxide per hundred grams of melt,

$$n_{100} = \frac{n_{Na_2O}}{100\,\text{g}} \tag{60}$$

Lux suggested that a pO scale be developed for oxide melts analogous to the pH scale:

$$\mathrm{pO} = -\log n \tag{61}$$

It may be noted that pO varies in the direction opposite from pH: a drop in pO accompanies an increase in melt basicity.

Flood and co-workers[29-31] showed that the relationship between the Brønsted–Lowry and Lux definitions is more than formal, citing as an example the fact that the proton-donating strengths of H_2SO_4 and H_2CO_3 parallel the oxide-accepting strengths of SO_3 and CO_2; that is, stronger Brønsted acids are related to stronger Lux acids.

Flood also drew an analogy between polyprotic acids and a similar phenomenon in oxide melts:

$$H_3PO_4 \overset{-H^+}{\rightleftharpoons} H_2PO_4^- \overset{-H^+}{\rightleftharpoons} HPO_4^{2-} \overset{-H^+}{\rightleftharpoons} PO_4^{3-} \tag{62}$$

$$SiO_2 \overset{+O^{2-}}{\rightleftharpoons} SiO_3^{2-} \overset{+O^{2-}}{\rightleftharpoons} SiO_4^{4-} \tag{63}$$

Potentiometric and indicator methods have been adapted to the determination of oxide activities. Chromate ion, whose decomposition to chromic ion at constant oxygen partial pressure is a function of oxide activity, serves as an oxide melt acidity–basicity indicator:

$$2CrO_4^{2-} \rightleftharpoons 2Cr^{3+} + 1.5O_2 + 5O^{2-} \tag{64}$$

Thermal decomposition and displacement reactions also furnish a qualitative idea of acid and base strength; for example,

$$MCO_3 \rightleftharpoons MO + CO_{2(g)} \tag{65}$$

$$MCO_3 + SO_{3(g)} \rightleftharpoons MSO_4 + CO_{2(g)} \tag{66}$$

Flood employed the former technique to obtain a qualitative order of oxide basicity, finding (as expected) that the most strongly basic oxides (those with the weakest conjugate acid metal ions) are those of the alkali and alkaline-earth metals. Flood obtained an order of acid strength in oxide melts from displacement reactions.

Flood also suggested that the Lux theory can be generalized to sulfide, fluoride, and other melts by simple analogy.

3.5 DONOR-ACCEPTOR ACIDS AND BASES AND IONOTROPY

In 1949 Ebert and Konopik[6,32,33] introduced a concept that differentiates between two kinds of acids and two kinds of bases in autoionizing solvents. Donor acids (A_D) are capable of releasing solvent cations or any other acid upon direct dissociation, and may alternatively manifest their acidity by reacting with the solvent to form solvent cations, for example,

$$HCl + H_2O \rightleftharpoons H_3O^+ + Cl^- \tag{67}$$

Donor bases (B_D) split off solvent anions or any other base in their dissociation, and may also react with the solvent to produce solvent anions,

$$NaOH_{(aq)} \rightarrow Na^+ + OH^- \tag{68}$$

Acceptor acids (A_A) are capable of binding solvent anions or any other base,

$$CO_2 + H_2O \rightleftharpoons HCO_3^- + H^+ \tag{69}$$

and acceptor bases (B_A) can bind solvent cations or any other acid,

$$NH_3 + H_2O \rightleftharpoons NH_4^+ + OH^- \tag{70}$$

Salt formation remains the characteristic neutralization process in this concept, but a comparison of reactions between the different types of acids and bases reveals a fundamental alteration in the role of the solvent compared to the classical neutralization process [equation (2)]. It is no longer necessary, according to this viewpoint, that the solvent always be a reaction product. On the contrary, it may be a product, reactant, or neither, depending on the reacting acid and base:

$$A_D + B_D \rightarrow \text{Salt} + \text{Solvent} \tag{71a}$$

for example,

$$HCl + NaOH \rightarrow NaCl + H_2O \tag{71b}$$

$$A_A + B_A + \text{Solvent} \rightarrow \text{Salt} \tag{72a}$$

for example, $$H_2O + 2NH_3 + SO_2 \rightarrow (NH_4)_2SO_3 \tag{72b}$$

$$A_A + B_A \rightarrow \text{Salt} \tag{73a}$$

for example, $$BCl_3 + (CH_3)_3N \rightarrow Cl_3BN(CH_3)_3 \tag{73b}$$

$$A_D + B_A \rightarrow \text{Salt} \tag{74a}$$

for example, $$HCl + NH_3 \rightarrow NH_4Cl \tag{74b}$$

$$A_A + B_D \rightarrow \text{Salt} \tag{75a}$$

for example, $$Zn(OH)_2 + 2NaOH \rightarrow Na[Zn(OH)_4] \tag{75b}$$

The Ebert–Konopik concept possesses two advantages over the Germann and Cady–Elsey definitions: it does not restrict acids and bases to substances linked to solvent ion concentration, and it recognizes the related fact that the solvent does not always participate in acid–base reactions, for example, equation (74b) in benzene. However, the donor–acceptor acid–base scheme is little more than a convenient method for classifying certain types of reactions and offers no theoretical foundation upon which to link donor and acceptor acids (or bases). It is therefore a step back in the search for the quality or qualities responsible for the characteristic properties of all acids and bases. There is no reason to create an artificial distinction between substances that behave identically, for example, towards indicators.

Gutmann and Lindqvist[6,32] pointed out that the transfer of ions from one species to another forms the basis for several acid–base theories. Proton transfer is the defining process in the Brønsted–Lowry theory; the Lux–Flood theory is based on oxide transfer; and Ebert and Konopik implicitly indicated that other ionic transfers are possible in acid–base reactions; for example, equations (69) and (75b) may be regarded as examples of hydroxide transfer. Gutmann and Lindqvist united and generalized these ideas into an acid–base theory founded on ion transfer, or "ionotropy," which they believed was crucial to the existence of acidity and basicity in ionizing solvents and in melts.

Ionotropic solvosystems are classified according to whether a cation or anion is transferred. In cationotropic solvents acids are cation donors and bases cation acceptors; cation transfer takes place from acid to base. In anionotropic solvents the direction of ion transfer is reversed: an acid is an anion acceptor and a base an anion donor.[6,32,34,35] The Gutmann–Lindqvist definitions may be summarized by

$$\text{Cationotropy: Acid} \rightleftharpoons \text{Base} + \text{Cation} \tag{76}$$

$$\text{Anionotropy: Base} \rightleftharpoons \text{Acid} + \text{Anion} \tag{77}$$

The ionotropic view of neutralization combines the classical form [equation (2)] with the double conjugate pair concept. Hence the products of a neutralization reaction remain a salt and pure solvent, but the ions of these products are recognized as having weak acid–base properties.

The only known cationotropic solvents are the prototropic solvents, that is, solvents in which autoionization involves proton transfer. On a practical basis cationotropy is therefore identical with the Brønsted–Lowry and Franklin concepts. Several types of anionotropic solvents and melts are included in the Gutmann–Lindqvist approach. The anion transferred may be fluoride, chloride, bromide, iodide, oxide, or sulfide.[5,6,22,32,35–37] Typical reactions in three different anionotropic media are shown in Table 3.2. Oxidotropy explains the exchange of oxygen between sulfite ion and sulfur dioxide solvent, in which autoionization is too limited to be responsible for the observed rapid exchange rate.[20,21] Gutmann and Lindqvist believed that their definitions could also be extended to solventless ion transfers.[6,32]

Potentiometric, conductimetric, and spectrophotometric methods have been adapted to acid–base studies in anionotropic solvents,[6,32,34,36] particu-

TABLE 3.2 Reactions in Selected Anionotropic Media

Fluoridotropy:
Solvent autoionization: $2BrF_3 \rightleftharpoons BrF_2^+ + BrF_4^-$
Acidic reaction: $SbF_5 + BrF_3 \rightleftharpoons SbF_6^- + BrF_2^+$
Basic reaction: $KF + BrF_3 \rightleftharpoons K^+ + BrF_4^-$
Neutralization: $[BrF_2][SbF_6] + KBrF_4 \rightarrow KSbF_6 + 2BrF_3$
Chloridotropy:
Solvent autoionization: $2AsCl_3 \rightleftharpoons AsCl_2^+ + AsCl_4^-$
Acidic reaction: $FeCl_3 + AsCl_3 \rightleftharpoons FeCl_4^- + AsCl_2^+$
Basic reaction: $C_5H_5N + 2AsCl_3 \rightleftharpoons C_5H_5NAsCl_2^+ + AsCl_4^-$
Neutralization: $[AsCl_2][FeCl_4] + [C_5H_5NAsCl_2][AsCl_4] \rightarrow [C_5H_5NAsCl_2][FeCl_4] + 2AsCl_3$
Bromidotropy:
Melt autoionization: $2HgBr_2 \rightleftharpoons HgBr^+ + HgBr_3^-$
Acidic reaction: $Hg(ClO_4)_2 + HgBr_2 \rightleftharpoons 2HgBr^+ + 2ClO_4^-$
Basic reaction: $NaBr + HgBr_2 \rightleftharpoons Na^+ + HgBr_3^-$
Neutralization: $[HgBr][ClO_4] + NaHgBr_3 \rightarrow NaClO_4 + 2HgBr_2$

larly by Gutmann and co-workers,[5,34,38,39] who employed these methods to provide experimental confirmation of ionotropy, using the solvent phosphorus oxychloride:

$$2POCl_3 \rightleftharpoons POCl_2^+ + POCl_4^- \tag{78}$$

or, for short,

$$POCl_3 \rightleftharpoons POCl_2^+ + Cl^- \tag{79}$$

where the chloride ion is implicitly solvated (just as the symbol H^+ in a protic solvent implies a solvated proton). Gutmann determined orders of acidic and basic strength in this solvent.

Ferric chloride is expected to manifest its acidity in $POCl_3$ via the characteristic chloridotropic reaction

$$FeCl_3 + POCl_3 \rightleftharpoons FeCl_4^- + POCl_2^+ \tag{80}$$

which is confirmed spectrophotometrically and by the yellow color of the tetrachloroferrate ion in dilute or alkaline solutions. However, the reddish-brown color of $FeCl_3$ is evident in concentrated or strongly acidic solutions and spectrophotometry yields no evidence of $FeCl_4^-$. X-ray diffraction shows that $FeCl_3$ is coordinated with the oxygen atom of a solvent molecule, and not with a chloride ion, in concentrated or strongly acidic $POCl_3$ solutions. Gutmann rationalized this behavior by postulating competition for the acid between chloride and oxygen sites in $POCl_3$:

$$Cl_3FeOPCl_3 \rightleftharpoons FeCl_4^- + POCl_2^+ \tag{81}$$

Solution acidity, basicity, and concentration determine the position of equilibrium in equation (81). The presence of a stronger acid, for example, $SbCl_5$, or of $FeCl_3$ in high concentrations shifts the equilibrium towards coordination of the ferric chloride with oxygen. Equation (81) is shifted toward $FeCl_4^-$ formation in alkaline or dilute $FeCl_3$ solutions because no stronger acid is present to compete for chloride ion. Gutmann compared this competition to an analogous situation in water: dilution of an aqueous solution of $FeCl_3$ in HCl leads to the gradual disappearance of yellow $FeCl_4^-$ as water (via oxygen) replaces chloride ion in the coordination sphere of the ferric ion.

Meek and Drago[5,40] criticized ionotropy, with particular reference to Gutmann's experiments in phosphorus oxychloride. They obtained spectrophotometric and titrimetric results for ferric chloride in triethylphosphate,

$(C_2H_5O)_3PO$, a solvent incapable of chloridotropy, that were similar to Gutmann's, and consequently suggested that solute dissociation is identical in both solvents, that is, the assumption of solvent ionotropy is incorrect and the solvent plays no role in dissociation:

$$2FeCl_3 \rightleftharpoons FeCl_2^+ + FeCl_4^- \qquad (82)$$

Gutmann[5] retorted that dichloroferric ion exists only in concentrated or strongly acidic $POCl_3$ solutions, but not in dilute ones, and that equation (82) is consequently valid only over a limited acidity range. Huheey[5,41] considered the controversy semantic and without foundation, declaring that circumstances in both solvents differ with regard to the possibility of leveling $FeCl_2^+$. Huheey's view is that triethylphosphate is unable to level the dichloroferric ion but that $POCl_3$ may be able to do so if its solvent cation is an acid weaker than $FeCl_2^+$:

$$FeCl_2^+ + POCl_3 \rightleftharpoons FeCl_3 + POCl_2^+ \qquad (83)$$

The lack of solute–solvent chloride exchange in phosgene[11,13] is also evidence against ionotropy, at least in this solvent.

3.6 ADVANTAGES AND DISADVANTAGES OF THE SOLVENT SYSTEMS THEORY

The unquestionable advantage of the solvent systems concept compared to the protonic concept is that the former extends acid–base processes to aprotic species.[1] There are, however, several problems associated with the theory. Defining acids and bases in terms of solvent cations and anions provides no information about general properties of acids and bases, for example, behavior toward indicators and catalysis. A great deal of reliance is placed upon ionization as a requisite property of acids and bases, and only lip service is paid to inherent acidity and basicity, that is, interactions, in the nonionized state[5] and reactions in the absence of a solvent.[1,7] Extension of the definitions into solvents like phosgene and sulfur dioxide, in which the identities of some of the solvent ions are uncertain,[5] leads to erroneous predictions. Finally, although acidities and basicities of protic solvents may be compared, there is no way of relating aprotic and protic acidity and basicity in the solvent systems theory.

REFERENCES

1. Day, Jr., M. C., and Selbin, J., "Theoretical Inorganic Chemistry," 2nd ed., Reinhold, New York, 1969.
2. Franklin, E. C., *J. Am. Chem. Soc.*, **27**, 820 (1905).
3. Franklin, E. C., *ibid.*, **46**, 2137 (1924).
4. Bell, R. P., *Q. Rev. Chem. Soc.*, **1**, 113 (1947).
5. Huheey, J. E., "Inorganic Chemistry, Principles of Structure and Reactivity," Harper and Row, New York, 1972.
6. Gyenes, I., "Titration in Nonaqueous Media," Van Nostrand, Princeton, New Jersey, 1967.
7. Hall, N. F., *J. Chem. Educ.*, **17**, 124 (1940).
8. McReynolds, J. P., *ibid.*, **17**, 116 (1940).
9. Germann, A. F. O., *Science*, **61**, 70 (1925).
10. Germann, A. F. O., *J. Am. Chem. Soc.*, **47**, 2461 (1925).
11. Huston, J. L., *J. Inorg. Nucl. Chem.*, **2**, 128 (1956).
12. Huston, J. L., and Lang, C. E., *ibid.*, **4**, 30 (1957).
13. Huston, J. L., *J. Phys. Chem.*, **63**, 383 (1959).
14. Cady, H. P., and Elsey, H. M., *Science*, **56**, 27 (1922).
15. Cady, H. P., and Elsey, H. M., *J. Chem. Educ.*, **5**, 1425 (1928).
16. Wickert, K., and Jander, G., *Ber.*, **70**, 251 (1937).
17. Wickert, K., *Z. Phys. Chem.*, **178**, 361 (1937).
18. Cruse, K., *Z. Elektrochem.*, **46**, 571 (1940).
19. Grigg, E. C. M., and Lauder, I., *Trans. Faraday Soc.*, **46**, 1039 (1950).
20. Norris, T. H., *J. Phys. Chem.*, **63**, 389 (1959).
21. Johnson, R. E., Norris, T. H., and Huston, J. L., *J. Am. Chem. Soc.*, **73**, 3052 (1951).
22. Gutmann, V., *Q. Rev. Chem. Soc.*, **10**, 451 (1956).
23. Lewis, G. N., "Valence and the Structure of Atoms and Molecules," Chemical Catalog Co., New York, 1923.
24. Ginell, R., *J. Chem. Educ.*, **20**, 250 (1943).
25. Wickert, K., *Z. Elektrochem.*, **47**, 330 (1941).
26. Smith, G. B. L., *Chem. Rev.*, **23**, 165 (1938).
27. Peterson, W. S., Heimerzheim, C. J., and Smith, G. B. L., *J. Am. Chem. Soc.*, **65**, 2403 (1943).
28. Lux, H., *Z. Elektrochem.*, **45**, 303 (1939).
29. Flood, H., and Forland, T., *Acta Chem. Scand.*, **1**, 592 (1947).
30. Flood, H., and Forland, T., *ibid.*, **1**, 781 (1947).
31. Flood, H., Forland, T., and Roald, B., *ibid.*, **1**, 790 (1947).
32. Gutmann, V., and Lindqvist, I., *Z. Phys. Chem.*, **203**, 250 (1954).
33. Ebert, L., and Konopik, N., *Oesterr. Chem. Ztg.*, **50**, 184 (1949).
34. Gutmann, V., *J. Phys. Chem.*, **63**, 378 (1959).
35. Gutmann, V., *Chem. Tech.* (*Leipzig*), **7**, 255 (1977).

36. Lindqvist, I., *Acta Chem. Scand.*, **9**, 73 (1955).
37. Gillespie, R., "Proton Transfer Reactions," Caldin, E., and Gold, V., Eds., Chapman and Hall, London, 1975. Chap. 1.
38. Gutmann, V., and Mairinger, F., *Monatsh. Chem.*, **89**, 724 (1958).
39. Gutmann, V., and Baaz, M., *Z. Anorg. Allg. Chem.*, **298**, 121 (1959).
40. Meek, D. W., and Drago, R. S., *J. Am. Chem. Soc.*, **83**, 4322 (1961).
41. Huheey, J. E., *J. Inorg. Nucl. Chem.*, **24**, 1011 (1962).

CHAPTER

4

LEWIS THEORY

Neither the protonic nor the solvent systems concepts provided an explanation for acid–base behavior observed for aprotic, nonionic solutes in aprotic, nonionizing solvents. Meyer[1] found that the colors and absorption spectra of quinone indicators in the presence of sulfuric, nitric, and hydrochloric acids dissolved in benzene or chloroform resembled the colors and spectra of the same indicators when $SnCl_4$, $ZnCl_2$, $SbCl_5$, $AlCl_3$, $HgCl_2$, $FeCl_3$, and SO_2 are dissolved in the same solvents. Folin and Flanders[2,3] noted, during their investigation into the necessity of an ionization prerequisite for the reaction of hydrogen acids and bases, that $HgCl_2$ in 50% ethanol–50% benzene solution can be titrated almost quantitatively to a phenolpthalein end point with sodium ethoxide. These observations, coupled with the mutual exclusiveness of certain aspects of the Brønsted–Lowry and solvent systems theories, indicated the need for a more inclusive acid–base concept.

4.1 LEWIS ACID–BASE DEFINITIONS

Lewis first published his ideas about acids and bases in 1923,[4] the same year in which the Brønsted–Lowry theory appeared. Lewis agreed with Brønsted that hydroxide ion is not the sole manifester of basic properties, but saw no reason to limit acids to hydrogen-containing compounds, pointing out the resemblance between reactions of protic and aprotic substances with bases[5]:

$$H^+ + OH^- \rightleftharpoons H_2O \tag{1}$$

$$Ag^+ + OH^- \rightleftharpoons AgOH \tag{2}$$

$$H_2O + NH_3 \rightleftharpoons NH_4OH \tag{3}$$

$$AgOH + NH_3 \rightleftharpoons H_3NAgOH \tag{4}$$

Lewis also objected to the definition of acids and bases on the basis of solvent cations and anions, respectively, according to the solvent systems theory because acids and bases are thereby restricted to species characteristic of ionizable (amphoteric) solvents, although acid–base phenomena are observed in nonionizing solvents of both the inert, for example, benzene, and reactive types, for example, pyridine[3,6,7]:

$$HCl + C_5H_5N \rightleftharpoons C_5H_5NH^+ + Cl^- \tag{5}$$

$$AsCl_3 + C_5H_5N \rightleftharpoons C_5H_5NAsCl_2^+ + Cl^- \tag{6}$$

It may be noted in equations (5) and (6) that no characteristic solvent cation is formed, whereas in an amphoteric solvent, such as water, a characteristic solvent cation is formed by solvent interaction with any acidic substance:

$$HCl + H_2O \rightleftharpoons H_3O^+ + Cl^- \tag{7}$$

$$AsCl_3 + 6H_2O \rightleftharpoons As(OH)_3 + 3H_3O^+ + 3Cl^- \tag{8}$$

Lewis believed, as did Brønsted, that ions were overemphasized as the carriers of acid–base properties, to the neglect of neutral molecules, despite the fact that many ions possess negligible acid–base properties, for example, perchlorate or tetraalkylammonium ions, and that, consequently, any acid–base definition formulated in terms of ions was incorrect and inadequate. Although fairly accurate calculations of heats of neutralization based on electrostatic attraction are possible, Lewis maintained that the consideration of acid–base processes as phenomena dependent on Coulombic forces is incorrect because these forces are predominant only at large interionic separation and are in part supplanted by other factors in a close-contact process like neutralization.[5]

Lewis proposed a generalized acid–base definition founded upon a mechanistic approach to chemical behavior instead of structure or constitution.[3–14] A base is a substance with a lone electron pair available to complete the stable electronic configuration (octet or other) of another atom, thereby increasing the stability of the latter. An acid employs a lone electron pair from another molecule to complete the stable electronic configuration of one of its own atoms. In other words, a base is an electron pair donor and an acid is an electron pair acceptor. Sidgwick[7,8,13,15,16] reached identical conclusions on the basis of Werner's coordination theory and formulated an acid–base (A–B) reaction as

$$A + :B \rightleftharpoons A:B \tag{9}$$

An important consequence of the early Lewis–Sidgwick definitions is the expansion of the acid category to include a wide range of substances not previously recognized as acids. Many of these, for example, SO_3, SiO_2, CO_2, BCl_3, and $SnCl_4$, had been considered acids prior to the advent of any theory, when acids were classified solely on the basis of experimental observation.[5–8] Sidgwick[10,15] classified electron pair acceptors into four categories:

1. Atoms with electron sextets, which are unstable and therefore are strong acids, for example, sulfur or oxygen:

$$:\ddot{\underset{..}{O}} + :N(C_2H_5)_3 \rightleftharpoons :\ddot{\underset{..}{O}}:N(C_2H_5)_3 \tag{10}$$

2. Compounds containing atoms with incomplete octets, for example, BCl_3:

$$\begin{array}{c} :\ddot{Cl}: \\ :Cl:B \\ :\ddot{Cl}: \end{array} + :NH_3 \rightleftharpoons \begin{array}{c} :\ddot{Cl}: \\ :\ddot{Cl}:\ddot{B}:NH_3 \\ :\ddot{Cl}: \end{array} \tag{11}$$

3. Metal ions with stable configurations, whose positive charge abets coordination:

$$Al^{3+} + 6:OH_2 \rightleftharpoons Al(:OH_2)_6^{3+} \tag{12}$$

4. Substances that already have a stable electronic configuration but are capable of expanding to larger ones:

$$\begin{array}{c} :\ddot{F}: \\ :\ddot{F}:Si:\ddot{F}: \\ :\ddot{F}: \end{array} + 2:\ddot{F}:^- \rightleftharpoons \left[\begin{array}{ccc} & :\ddot{F}: & \\ :\ddot{F}: & & :\ddot{F}: \\ & Si & \\ :\ddot{F}: & & :\ddot{F}: \\ & :\ddot{F}: & \end{array} \right]^{2-} \tag{13}$$

It may be noted that these categories sometimes overlap. Sidgwick included hydrogen acids in the fourth group, since he believed that hydrogen expanded its stable configuration from two to four electrons upon reaction with a base, forming two-covalent hydrogen:

$$ClH + :OH_2 \rightleftharpoons ClH:OH_2 \rightleftharpoons Cl^- + H:OH_2^+ \tag{14}$$

$$HOH + :NH_3 \rightleftharpoons HOH:NH_3 \rightleftharpoons OH^- + H:NH_3^+ \tag{15}$$

The bond holding the newly coordinated proton in equations (14) and (15) is indistinguishable from those previously present in the cations formed, and thus $H:OH_2^+$ is written as H_3O^+, and $H:NH_3^+$ as NH_4^+.

A fifth category may be added:

5. Multiply bonded acidic center compounds that shift one of their own electron pairs to accomodate a simultaneously entering electron pair. Carbon dioxide is an example of this group:

$$:\ddot{O}::C::\ddot{O}: + :OH^- \rightleftharpoons HO:CO_2^- \qquad (HCO_3^-) \qquad (16)$$

Lewis[5,6] believed that all elements, in one form or another, are capable of acidic behavior, except for the heavier alkali and alakline-earth metals and the rare gases. Bases are restricted to substances containing elements in periodic groups V A–VII A (and sometimes carbon); Lewis bases are thus essentially identical to Brønsted bases.

The initial publication of these ideas[4] did not attract much attention, and Lewis theory was virtually ignored for 15 years by the bulk of the scientific community, mainly because it represented a radical departure from accepted views of the nature of chemical bonding, as well as a new view of acids and bases. The theory of covalent bonding, also developed largely by Lewis, was only seven years old at the time his acid–base concept first emerged, and scientists who still held to purely ionic bonding theories (outside of Werner's coordination complexes) were reluctant to accept any theory predicated on covalence.[13] The advent of the valence bond and molecular orbital theories in the 1930s, in which all bond types were regarded not as innately different but as different degrees along a continuum, led to increased acceptance of covalent bonding and, subsequently, of Lewis acid–base theory. Another group of traditionalists objected to any theory that removed hydrogen from its century-old position as the "acidifying principle." A partial reason for the earlier neglect of these ideas lay in the manner in which Lewis first presented them.[4,13] They were merely stated in passing, without elaboration or supporting experimental evidence. When Lewis again put forth his views in 1938[5] he included experimental evidence to support his definitions.

The supporting evidence cited by Lewis is divided according to four criteria considered to be definitive acid–base characteristics: neutralization, displacement, indicator reactions, and catalysis.[3,5–7,12–14,16–18]

1. *Neutralization* The typical acid–base reaction is rapid neutralization [equation (9)], which, in the Lewis sense, connotes completion of the stable electronic configuration of the acceptor atom of the acid by an electron pair (or pairs) from the base. The covalent bond thus formed differs from ordinary covalent bonds in its manner of formation, that is, both bonding electrons in the former case are supplied by the base [equations (9)–(16)], whereas ordinary covalent bonds require each reactant to provide one of the bonding electrons, for example,

$$2:\underset{..}{\ddot{Cl}}\cdot \rightleftharpoons :\underset{..}{\ddot{Cl}}:\underset{..}{\ddot{Cl}}: \tag{17}$$

The product of a Lewis acid–base reaction is called a coordination compound, coordination complex, or adduct.[10] Neither salt nor conjugate acid–base formation is a requirement of neutralization, although some acid–base processes produce an electrical strain that causes dissociation or rearrangement to accompany the neutralization process, for example, alteration of the geometries of BCl_3 and SiF_4 in equations (11) and (13), respectively, or dissociation of the adducts formed in equations (14) and (15).[8,9]

The predecessors of the electronic theory, with the notable exception of Hantsch's pseudoacid concept, never explicitly or implicitly considered the speed of neutralization to be of any importance, because all the acid–base reactions recognized by the Arrhenius, Brønsted–Lowry, and solvent systems theories are instantaneous. This is not true for all Lewis neutralization processes, some of which are extremely slow. For example, Folin and Flanders[2] inferred that carbon dioxide was not an acid outside aqueous solution, since it could not be titrated in benzene or chloroform; it may noted that these solvents are inert (nonleveling), so that the lack of apparent acidity on the part of CO_2 is not due to a leveling of its strength to that of a solvent cation. Lewis recognized and addressed this problem in the second publication of his acid–base ideas, attempting to integrate slow neutralization processes into the electronic concept by distinguishing between "primary" and "secondary" acids and bases.[3,5,7,9] Primary acids and bases possess electronic configurations conducive to direct combination [equations (9)–(13)]; no structural alteration nor activation energy is required for the reaction of a primary acid with a primary base. Secondary acids and bases are not reactive in the state of lowest energy and cannot combine directly, but they become reactive if sufficient activation energy is supplied to cause internal excitation and changes in electronic structure. Solvation is one way of supplying the required activation energy if the solvent is able to neutralize the acid or base; the accompanying shift in electron density makes the solute reactive towards neutralization by a stronger base or acid, as appropriate. Lewis stated that some acid–base reactions occur only in the presence of a suitable solvent. For a secondary acid the effect of solvation is represented by amending equation (9):

$$A:B' + :B \rightleftharpoons A:B + :B' \tag{18}$$

where B′ is a basic solvent. Analogously, an acidic solvent A′ may render a secondary base reactive:

$$A':B+A \rightleftharpoons A:B+A' \tag{19}$$

Excitation energy may also activate a secondary acid or base by promoting a shift in electron density through bond breakage, as in the case of CO_2 [equation (16)], which Lewis believed becomes reactive by breaking a bond and producing an acidic carbon atom:

$$\underset{\text{Secondary acid}}{:\ddot{O}::C::\ddot{O}:} + E_{act} \rightleftharpoons \underset{\text{Primary acid}}{:\ddot{O}::\overset{+}{C}:\overset{-}{\ddot{O}}:} \tag{20}$$

where E_{act} represents the activation energy. Equation (20) indicates that electronic shifts render secondary species reactive by changing them into primary species. Lewis therefore postulated the existence of electromers for secondary acids and bases.[19,20] Electromerism is an analogue of isomerism in which the atoms of a molecule retain their relative positions and only electrons shift in arrangement, for example, equation (20). Electromerism differs from resonance, according to Lewis, in that electromers, like isomers, possess different physical and chemical properties, whereas resonance structures do not. For example, the neutralization product of the ground-state (secondary) electromer of the tris(4-nitrophenyl)methide ion with hydrogen acids is colorless,

$$(p\text{-}O_2NC_6H_4)_3C:^- + H^+ \rightleftharpoons (p\text{-}O_2NC_6H_4)_3CH \tag{21}$$

while that of its excited (primary) electromer is orange[20]:

$$(p\text{-}O_2NC_6H_4)_2C{=}C_6H_4{=}NO_2^- + H^+ \rightleftharpoons (p\text{-}O_2NC_6H_4)_2C{=}C_6H_4{=}NOOH \tag{22}$$

Secondary acids and bases may also be identified by the noticeable drop in neutralization rates with decreasing temperature (decreasing activation energy). Primary acid–base reactions remain instantaneous regardless of temperature.[19]

Triphenylmethide ion derivatives are the only bases described as secondary by Lewis. Secondary acids include, in addition to CO_2, triphenylmethylcarbonium ion and its derivatives, acid anhydrides, and nitro-organics, for example, nitromethane. Objections to the existence of two-covalent hydrogen [equations (14) and (15)], to be discussed in detail later, caused Lewis to also classify hydrogen acids as secondary acids. According to this viewpoint, all hydrogen acids contain the primary acid H^+. The typical Brønsted–Lowry acid–base reaction may be represented by equation (18), in which A is the proton and B′ is no longer necessarily the solvent. The secondary acid category is therefore stretched to include substances that are strictly Lewis acid–base adducts, that is, hydrogen acids. There is a basis for this, since the proton does not exist in the free state, and therefore direct acid–base reaction via equation (9) is not possible.

Later recognition that the majority of Lewis acid–base processes take place via equations (18), (19), or

$$A:B' + A':B \rightleftharpoons A:B + A':B' \tag{23}$$

instead of direct combination, rendered the distinction between primary and secondary acids and bases obsolete, but its remarkable resemblance to the pseudo–true acid theory of Hantsch may be noted.

2. *Displacement* Weak acids are displaced from adducts by strong acids, and weak bases by strong bases, for example, the basic displacement typified by equation (18) and the acidic displacement characterized by equation (19). Brønsted–Lowry protolyses are conceived of, in Lewis' view, as reactions in which a strong base displaces a weaker base from coordination with the proton,[10] for example, the replacement of chloride ion by water in equation (14). The reaction of BCl_3 or $SnCl_4$ with sodium carbonate in mixed CCl_4–acetone solvent is an example of acid displacement. The weak acid CO_2 is evolved and the base sodium oxide forms a coordination complex with the stronger acid. Double displacements [equation (23)] occurring between two coordination compounds are simultaneous acid and base displacements.

3. *Indicator reactions* The common indicator methods of measuring acidity, basicity, and the degree of neutralization are applicable to Lewis acids and bases. The color changes of a given indicator often vary with the solvent, but indicator color is independent of the nature of solute acids and bases in a given solvent; that is, it is immaterial whether the acid is protonic or aprotic. Lewis and Bigeleisen,[21] working along the same lines as Meyer,[1] found the colors and spectra of various indicators in the presence of BCl_3, $SnCl_4$, and hydrogen acids to be almost identical. It may be noted at this point that traces of Lewis acid impurities sometimes react with indicators to produce colors leading to incorrect estimations of acidity. For instance, a small amount of metal ion impurity present in a neutral or slightly alkaline aqueous solution may react with an acid–base indicator and convert it to its acid form, giving the solution a color characteristic of an acidic solution of the indicator.

4. *Catalysis* Lewis acids and bases catalyze inorganic and organic reactions. Acidic catalysts generally displace a Lewis base from one of the reactants, leaving a positively charged, electrophilic (acidic) fragment.[22] Occasionally the catalyst simply combines with a reactant to shift its electron density so that an electron deficiency is localized on a particular reactant atom, which then serves as the reaction site. For example, Hubbard and Luder[23] noted that the reaction of metals with thionyl chloride to produce metallic chlorides, sulfur, and SO_2 is catalyzed, in increasing order, by $SnCl_4$, $FeCl_3$, and $AlCl_3$. The reaction is believed to proceed via formation of the thionyl (SO^{2+}) ion, and the catalysis by Lewis acids is explained by the displacement of chloride from the reactant to the catalyst, thereby increasing the thionyl concentration:

$$2AlCl_3 + SOCl_2 \rightleftharpoons 2AlCl_4^- + SO^{2+} \tag{24}$$

The mechanism of the Friedel–Crafts alkylation of benzene by alcohols offers another example of Lewis acid catalysis, with BF_3 serving as the catalyst:

$$CH_3CH_2OH + BF_3 \rightleftharpoons CH_3CH_2^+ + F_3BOH^- \tag{25}$$

$$CH_3CH_2^+ + C_6H_6 \rightleftharpoons C_6H_5(H)^+CH_2CH_3 \tag{26}$$

$$C_6H_5(H)^+CH_2CH_3 \rightleftharpoons H^+ + C_6H_5CH_2CH_3 \tag{27}$$

$$H^+ + F_3BOH^- \rightleftharpoons F_3BOH_2 \tag{28}$$

The catalyst is recovered through dehydration. The mechanism is no different if a hydrogen acid, for example, HF, is employed as a catalyst instead of an aprotic Lewis acid.

Basic catalysis occurs analogously: the catalyst displaces an acid from a reactant, leaving a negatively charged, nucleophilic (basic) fragment, or the catalyst may simply combine with the reactant, shifting its electron density to localize an electron excess on a particular reactant atom, which then serves as the reactive site.[22]

Not all acid-catalyzed and base-catalyzed reactions are catalyzed by all Lewis acids and bases, respectively. Many require a specific catalyst; for example, the reaction between sulfur dioxide and alkoxide ions to produce sulfonate ions is catalyzed specifically by iodide and thiocyanate ions.[5] The polymerization of isobutene is not catalyzed by aprotic Lewis acids unless a trace amount of hydrogen acid is present as a co-catalyst to initiate the reaction.[17] A comprehensive survey of Lewis acid-catalyzed and base-catalyzed organic reactions has been given by Luder and Zuffanti.[3,22]

Lewis' inclusion of experimental criteria, coupled with increasing acceptance of covalence and parallel developments in the solvent systems theory during the 1930s and 1940s, resulted in the electronic acid–base theory becoming the basis for correlation of a great deal of supposedly unrelated chemistry.

Lewis' definitions are formulated without special consideration of solvents; that is, solvents are classified as Lewis acids or bases, according to their electron-pair-accepting and donating tendencies, just as other substances are. Early opponents of the Lewis concept feared that its acceptance would destroy the significance of dissociation constants and indicator measurements in a given solvent.[24] Luder and Zuffanti[3,7] believed such fears to be groundless, pointing out that the significance of equilibrium constants and indicator measurements is enhanced and clarified, rather than diminished, by considering the solvent as a reference Lewis acid or base.

The electronic theory recognizes three solvent classes: [3,6,7] inert, ionizing, and nonionizing reactive. One may consider the effect of each type of solvent upon a neutralization process taking place between a solute acid and solute base, for example, BCl_3 and triethylamine, to illustrate the distinction between the three classes:

1. Inert solvents, for example, benzene, take no part in the neutralization process, serving only as dilutents. Acid–base combination is direct:

$$Cl_3B + N(C_2H_5)_3 \rightleftharpoons Cl_3BN(C_2H_5)_3 \qquad (29)$$

2. Ionizing (amphoteric) solvents, for example, water or selenium oxychloride, behave as bases towards Lewis acids, and as acids towards Lewis bases, forming solvent ions and intermediate displacement products with both. Combination of an acidic solution with a basic solution yields the adduct of equation (29) and pure solvent:

$$BCl_3 + SeOCl_2 \rightleftharpoons BCl_4^- + SeOCl^+ \qquad (30)$$

$$SeOCl_2 + N(C_2H_5)_3 \rightleftharpoons (C_2H_5)_3NSeOCl^+ + Cl^- \qquad (31)$$

$$BCl_4^- + (C_2H_5)_3NSeOCl^+ + SeOCl^+ + Cl^- \rightleftharpoons$$

$$Cl_3BN(C_2H_5)_3 + 2SeOCl_2 \qquad (32)$$

The ability to behave either as an acid or as a base also accounts for the high degree of association in these solvents and their consequent physical properties, for example, high boiling point.

The solvents of this class are the only ones in which the solvent systems theory, predicated on the existence of solvent cations and anions, is valid. Luder and Zuffanti[3,7] emphasized the fact that the amphoterism of most of the common solvents, for example, water, alcohol, and ammonia, is the reason for the misconception of general applicability of the solvent systems concept. They also contended that many so-called "typical" acid–base properties, such as the ability of acidic solutions to dissolve metals and thereby produce gases characteristic of solvent cations, for example, H_2 and CO, are actually properties of solvent ions. According to Luder and Zuffanti, the dissolution of metallic magnesium in an aqueous acid solution depends not on the identity of the acid but on the hydronium ion concentration. Any acid dissolved in water increases the H_3O^+ concentration, be it protonic like HCl or aprotic like SO_3 and BCl_3. However, no reaction takes place between acids and metals in the inert solvent benzene because no solvent ions exist. Luder and Zuffanti therefore rejected acid–base properties apparent only in ionizing solvents as valid criteria for acid–base classification; they emphasized the more general experimental criteria proposed by Lewis.

3. Nonionizing reactive solvents, for example, pyridine or ethyl ether, comprise a class that is intermediate between inert and amphoteric solvents. This class is, in principle, an expanded version of Bronsted's protophilic and protogenic solvent categories. A nonionizing reactive solvent is capable of forming an intermediate displacement product with either the solute acid or base, but not with both:

$$Cl_3B + NC_5H_5 \rightleftharpoons Cl_3BNC_5H_5 \tag{33}$$

$$Cl_3BNC_5H_5 + N(C_2H_5)_3 \rightleftharpoons Cl_3BN(C_2H_5)_3 + C_5H_5N \tag{34}$$

The adduct formed is identical, regardless of the nature of the solvent [equations (29), (32), and (34), assuming that both the solute acid and base are stronger than the solvent], but of all the acid–base theories advanced up to this point, only the electronic theory is valid in circumstances where the solvent is neither amphoteric nor protogenic and the solute is aprotic.

Equation (33) also serves to illustrate the general validity of Hantsch's leveling effect when applied to aprotic processes that occur in nonionizing solvents. A strong acid is thus not necessarily leveled to a cation in a basic solvent, and a strong base is not always leveled to an anion in an acidic solvent, although ions are frequently the products of leveling. The electronic theory interpretation of the leveling effect also clarifies the ambiguous position assigned to metal ions and aquometal ions in the Brønsted–Lowry theory by regarding the latter, for example, $Al(H_2O)_6^{3+}$, as products of the leveling of stronger Lewis acids, for example, Al^{3+}, by water [equation (12)].

An additional, indirect effect to be considered as part of the solvent influence on Lewis acid–base processes is the action of the solvent in promoting or inhibiting the ionization that sometimes accompanies or follows the formation of a coordinate covalent bond. It was mentioned, in the discussion of Lewis' experimental acid–base criteria, that electrical strain is often a byproduct of neutralization. The dielectric constant of the solvent may augment or oppose this strain; a high dielectric constant favors dissociation, and a low one inhibits it.[16]

Amphoterism is not restricted to solvents by the electronic theory, but it is recognized as a property of most acids and bases.[2,3,5,7,16] For instance, HCl, which contains three unshared electron pairs, behaves as a base towards the stronger acids $SnCl_4$ and SO_3:

$$SnCl_4 + 2HCl \rightleftharpoons H_2SnCl_6 \tag{35}$$

$$SO_3 + HCl \rightleftharpoons ClSO_3H \tag{36}$$

In sulfur trioxide the sulfur atom is strongly acidic and the oxygen atoms are weakly basic; in HCl the hydrogen atom is acidic, and the chlorine atom basic. Equation (36) can therefore be rewritten to illustrate the amphoteric tendencies of both reactants:

$$\begin{array}{ccccccc} & :\ddot{O}: & & & & :\ddot{O}: & \\ :\ddot{O}: & S & + & :\ddot{Cl}: & \rightleftharpoons & :\ddot{O}:S:\ddot{Cl}: & \\ & :\ddot{O}: & & H & & :\ddot{O}:\ H & \end{array} \tag{37}$$

Developments in the solvent systems theory in the late 1930s and early 1940s paralleled the Lewis conception in emphasizing electronic configuration as a determinant in acid–base classification, but did not offer as comprehensive a treatment of as varied a group of chemical phenomena. Wickert[3,11,25,26] defined acids as substances dissociating into cations that lack closed-shell configurations and anions with closed-shell configurations; bases were defined as substances dissociating into cations and anions, both possessing closed-shell configurations. Implicit in these definitions are the potential capabilities of acids to accept electrons and of bases to donate electrons, but Wickert did not emphasize this aspect, choosing instead to attribute importance to ionization as a characteristic acid–base property. Wickert also claimed that bases are inherently linked to salts, and acids to "water-like" solvents, a restricted group of media outside of which his concept has very limited validity. Finally, Wickert gave amphoterism short shrift, and his definition of acids excludes many substances with acidic properties.

Cruse[27] regarded acids as species able to bind bases by accepting electrons, but considered this "electron exchange" process an aprotic analogue of protolysis instead of a more extensive phenomenon. The solvent systems definition closest to the electronic theory definition was enunciated by Smith,[11,28] who defined acids and bases as neutral or charged electron pair acceptors and donors, respectively, towards ions or molecules of a solvent, a qualification that restricts his definitions to ionizing solvents.

Gutmann and Lindqvist[29] contended that their ionotropic (ionic transfer) theory of acids and bases is fundamentally in agreement with the electronic theory, the difference being the frame of reference from which an acid–base reaction is viewed.

Lewis theory emphasizes coordinate covalent bond formation and ionotropy stresses ionization,[30] for example, in the solvent $AsCl_3$,

$$C_5H_5N + AsCl_3 \underset{\text{base reaction}}{\overset{\text{Lewis acid–}}{\rightleftharpoons}} C_5H_5NAsCl_3 \underset{\text{base reaction}}{\overset{\text{Ionotropic acid–}}{\rightleftharpoons}} C_5H_5NAsCl_2^+ + Cl^- \tag{38}$$

Gutmann and Lindqvist also believed that the major fault of Lewis theory lies in the fact that it does not recognize the special influence of solvents on acid–base processes.[29]

Tracer studies in phosgene and in sulfur dioxide[31,32] pointed to the apparent lack of solvent ionization in the former and the inability to characterize a definite solvent cation in the latter medium as evidence favoring the applicability of the electronic theory over ionotropy, at least in these solvents. Ionotropy is considered as a possibility only in solvents in which definite solvent cationic and anionic entities are known. Meek and Drago[33,34] went even further, questioning the existence of ionotropy. Having obtained spectrophotometric and titrimetric results for acidic ferric chloride in the nonchloridotropic solvent triethylphosphate that were almost identical to those obtained by Gutmann for the same solute in the supposedly chloridotropic solvent phosphorus oxychloride, they proposed that the interactions in both solvents are identical; since ionotropy is not possible in triethylphosphate, a Lewis mechanism must be operative in these and all (except inert) solvents.

The Lewis concept was also employed to interpret reactions in aprotic melts.[35] Lewis acids in such media include coordinately unsaturated ions, for example, metaphosphate (PO_3^-) and metaborate (BO_2^-), and macromolecules with coordinately unsaturated individual units, for example, silica and titania; oxide, fluoride, and sulfide ions are Lewis bases in melts.

The foregoing discussion included solvent systems theories insofar as they relate to the electronic theory. A more detailed discussion of solvent systems concepts in their own right appears in Chapter 3.

The electronic theory of acids and bases includes all substances and concepts covered by the Arrhenius, Brønsted–Lowry, and solvent systems theories, as well as species and ideas not included in any of the other concepts.[3,6,7,9,16,17] Lewis' approach incorporates coordination chemistry into acid–base chemistry and explains the observation that electron-withdrawing substituents increase the acidity and decrease the basicity of organic species, whereas electron-releasing substituents have the opposite effect. Even resonance structures may be interpreted by Lewis theory as cases of intramolecular neutralization[5]:

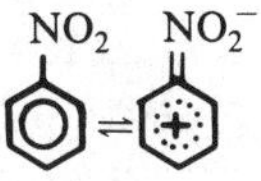

NO_2 NO_2^- (39)

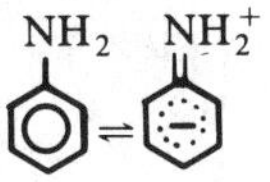

NH_2 NH_2^+ (40)

4.2 DISADVANTAGES OF THE EARLY LEWIS FORMULATION

Lewis' theory was criticized for several reasons, but mostly because of its generality. Critics argued that the concept practically equated "acid" and "base" with "reactant," and that the theory's correlation of a wide range of chemical phenomena, although valid, did not justify referring to acids and bases in such a general sense.[13,14,36]

A major point of contention was the position accorded the hydrogen acids by the electronic theory. Sidgwick,[15] in the early formulation of the concept, postulated doubly coordinated hydrogen in order to include hydrogen acids in the electron-pair-acceptor category. Lewis assumed that such coordination produces a large degree of electrical strain, forcing ionization of the acid–base adduct [equations (14)–(15)]. However, others[6,12,14,36,37] pointed out that the first step in the reaction of hydrogen acids with bases involves the formation of an intermolecular hydrogen bond that may be unstable and consequently undergoes immediate rearrangement, that is, transfer to the base, before any coordination occurs. The intermolecular hydrogen bond is a largely electrostatic, rather than covalent, entity that cannot be equated with the coordinate covalent bonds of Lewis theory. Therefore, since the initial nature of the bond between a base and a Bronsted acid differs substantially from that of the bond between a base and a Lewis acid, the postulation of two-covalent hydrogen is incorrect and the validity of the electronic theory for hydrogen acids (molecular or ionic, for example, NH_4^+), as well as for any acids lacking electron deficiencies, is in doubt. In fact, many Bronsted acids, for example, the hydrogen halides, are more easily classified as bases in the strict Lewis sense, because they contain available electron pairs but no electron deficiencies. An alternate interpretation of this limitation of the electronic theory regards hydrogen acids not as true acids but, rather, as adducts of the acid H^+ and a base.[14,17]

Advocates of the Lewis approach responded to these arguments by advancing a slightly different view, regarding Brønsted acids as proton–base adducts. According to this viewpoint, hydrogen acids are secondary acids that contain the primary acid H^+. The characteristic reaction of these acids with bases is a Lewis base displacement [equation (18)]. Luder[38] contended that the stress placed by the Brønsted theory on protolysis (Lewis base displacement) emphasizes displacement to the neglect of Lewis' three other identifying acid–base criteria. Opponents of the Lewis concept did not accept this viewpoint and continued to insist on a distinction between Brønsted and Lewis acids to the extent of proposing other terms, such as "protoacid," "secondary acid," "antibase," "pseudoacid," and "acceptor" to describe the latter, while retaining "acid" for hydrogen acids.[14,36,37,39]

A related argument concedes that Bronsted acids comprise a subclass of Lewis acids but distinguishes between Bronsted acidity (proton-donating tendency) and Lewis acidity (electron-pair-accepting tendency).[40] A consideration of the hydrogen halides reveals that their Bronsted acidities increase in the order $HF < HCl < HBr < HI$, but the decreasing electronegativity along this series indicates that the lighter halides impart more ionic character to the H–X bond and therefore Lewis acidity increases in the opposite direction; that is, HF is a better acceptor of bases, especially F^-, than the other hydrogen halides. Luder[41] believed this argument to also be a consequence of what he termed the "Bronsted theory's excessive preoccupation with displacement." The reversibility and rapidity of almost all protonic acid–base reactions (barring those involving intramolecular rearrangement as postulated by Hantsch) has also been contrasted with Lewis acid–base reactions, some of which are neither rapid nor reversible.[42]

Probably the most frequently cited criticism of the Lewis theory is the inability to arrange acids (or bases) into a single order of strength. This problem and attempts to deal with it will be discussed in detail in Section 4.4.

4.3 QUANTUM MECHANICAL FORMULATIONS OF THE LEWIS THEORY

Lewis[5] recognized that his emphasis on electronic structure as the definitive acid–base trait made his theory the only acid–base concept amenable to quantum mechanical treatment, but he did not undertake such a treatment himself. Mulliken[13,43–45] was the first to formulate the electronic theory in a quantum mechanical context, developing a general molecular orbital (MO) concept from an attempt to predict the spectral absorption bands and observed colors of charge-transfer complexes, a class of weak Lewis acid–base adducts, for example, iodine in water.

The wave function Ψ_{AB} of a 1 : 1 adduct may be represented as[13,43–45]

$$\Psi_{AB} = a\Psi(A, B) + b\Psi(A^{-}B^{+}) \tag{41}$$

where $\Psi(A, B)$ is a so-called "no-bond" wave function accounting for all electrostatic interactions between A and B, for example, ionic, permanent dipole, and induced dipole attractions. No electron transfer from B to A has taken place in the $\Psi(A, B)$ state. $\Psi(A^{-}B^{+})$ is the wave function of the system after complete transfer of one electron from B to A. The actual degree of electron transfer is intermediate between the pure $\Psi(A, B)$ and $\Psi(A^{-}B^{+})$ states in most adducts and is determined by the square of the ratio of the weighting coefficients a and b; b^2/a^2 varies from zero (no electron transfer) to infinity (complete electron transfer), that is, all degrees of donation are possible. The previous statement is one of the most important results of the quantum mechanical treatment because it provides a clear notion of the partial electron transfer that serves as the basis for coordinate covalent bond formation in Lewis' original formulation of the electronic acid–base theory. Mulliken noted, with regard to this aspect of his formulation, that the Bronsted–Lowry conception of proton transfer is reduced to a mere formality, since the actual interaction between a protic acid and a base involves a decrease in the degree of electron transfer to the proton from the weaker base and a simultaneous increase in the degree of electron transfer to the proton from the stronger base; that is, the proton is never free, not even at the instant of transfer.

Application of perturbation theory to equation (41) yields an expression of the energy of an acid–base complex in terms of the sum of electrostatic and charge-transfer (covalent) contributions[13]:

$$E_{AB} = E_0 - \frac{(\beta_{01} - E_0 S_{01})^2}{E_1 - E_0} \tag{42}$$

where E_{AB} is the energy of the A:B adduct, E_0 is the energy of $\Psi(A,B)$, including the energies of the isolated acid and base, as well as the energy of electrostatic interactions, E_1 is the energy of the $\Psi(A^-B^+)$ state, β_{01} is the resonance integral between the two states, and S_{01} is the overlap integral.

The first term of equation (42) is a function of the net charge densities of the donor and acceptor sites. The second term depends on the valence states of the interacting species, the ionization potential (IP) of the donor MO (preferably small), the electron affinity (EA) of the acceptor MO (preferably large), and the extent of orbital overlap. To a reasonable approximation,

$$E_1 - E_0 \simeq |IP|_B - |EA|_A \tag{43}$$

The numerator of the second term of equation (42) also includes orientational geometric factors contributing to a lowering of the activation energy for adduct formation.

The Mulliken formulation differs from Lewis' earlier conception in that the MO treatment of the former requires no localization on a particular atom of electron density excess or deficiency for a species to behave as a base or acid, respectively. Hence the interaction of iodine with the π-electron ring of benzene may be regarded as an acid–base process. Mulliken proceeded to classify all donor–acceptor reactions according to electrostatic (charge) and covalent (σ, πMO) donor–acceptor properties, producing a systematized and expanded Lewis approach to reactivity.

Analogous results were obtained by Klopman and Hudson,[13,46–48] who based a quantum mechanical treatment of reactivity between atoms in donor and acceptor species on an estimation of the degree of perturbation of reactant ground-state MOs Ψ_A and Ψ_B caused by the interaction of A with B:

$$\Psi_{AB} = a'\Psi_A + b'\Psi_B \tag{44}$$

where a' and b' again represent weighting coefficients. The Klopman–Hudson approach distinguishes between two types of interaction:

1. The neighboring or ionic effect is comprised of long-range electrostatic interactions, but not electron transfer. The former are functions of the charge densities q_s and q_t of the donor and acceptor atoms s and t of B and A, respectively; of the distance between these atoms, R_{st} (which is dependent on the radii of A and B); and of the solvent dielectric constant ε.

2. The effect of partial charge transfer increases the covalent nature and decreases the ionicity of the forces between A and B. This is accom-

plished by short-range MO overlap and depends on the individual MO symmetry and overlap properties, as well as the orbital energies of A and B as modified by solvation.

The total energy change on adduct formation, ΔE, is the sum of the ionic and covalent contributions:

$$\Delta E = -\frac{q_s q_t e^2}{R_{st}\varepsilon} + 2\sum_m \sum_n \frac{(c_s^m c_t^n \beta_{st})^2}{E_m^* - E_n^*} \tag{45}$$

where e is the electronic charge, β_{st} is the resonance integral, c_i^j is the orbital coefficient of the jth MO at atom i, E_m^* is the energy of the MO m of B in the field of A, corrected for solvation or desolvation accompanying the loss of an electron, and E_n^* is the energy of the MO n of A in the field of B, corrected for solvation or desolvation accompanying the gain of an electron. The first summation is over the occupied orbitals m of B, the second over the unoccupied orbitals n of A, and the factor of 2 indicates that two electrons are transferred in a Lewis acid–base interaction. Equation (45) may be simplified by assuming that covalent interactions are dominated by the highest occupied molecular orbital (HOMO) of the base (donor) and the lowest unoccupied molecular orbital (LUMO) of the acid (acceptor); the summations of equation (45) are thus reduced to a single term. The resemblance of the HOMO–LUMO concept to Lewis' classical formulation may be noted. The HOMO and LUMO are referred to as "frontier" orbitals.

The relative magnitudes of the numerator and denominator of the last term of equation (45) determine the extent of perturbation and the type of reactivity. When $2(c_s^m c_t^n \beta_{st})^2 \ll (E_m^* - E_n^*)$, the Coulombic term of equation (45) is predominant and the acid–base interaction is said to be "charge controlled." Very little charge transfer occurs, and the MO perturbation is small. Charge-controlled processes are favored by highly polar acceptors and donors, in which the reactive atoms have high positive and negative charge densities, respectively; by small interacting species; and by very small intermolecular distances. These factors increase the magnitude of the Coulombic term in equation (45). Circumstances that favor charge control by decreasing the degree of covalent bonding include situations in which the donor base is highly electronegative (high IP) and the acceptor acid has a low EA [see equation (43)], that is, $E_m^* - E_n^*$ is large; in these cases the stability gained upon charge transfer is small, and covalent interaction is therefore not favored (Figure 4.1*a*). A low degree of overlap (small β_{st}) also favors charge control.

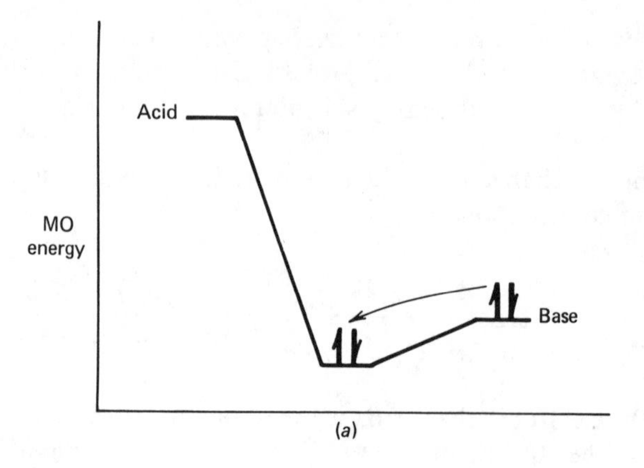

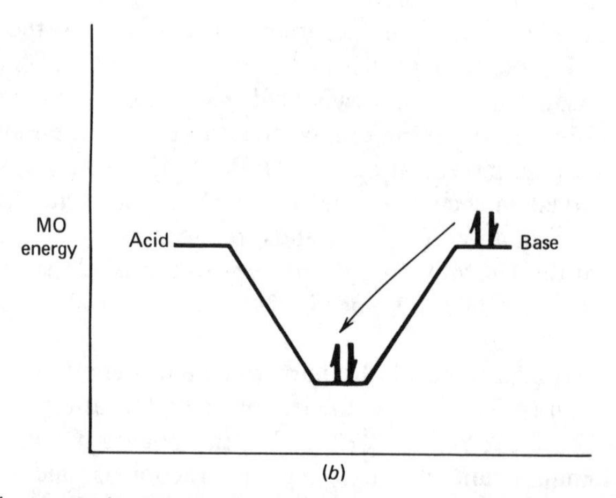

Figure 4.1. Charge transfer (covalent interaction) in charge-controlled and frontier-controlled processes. (a) In the charge-controlled process the stability gained upon charge transfer is small; hence covalent interaction is not favored. (b) In the frontier-controlled process the stability gained upon charge transfer is large; hence covalent interaction is favored.

On the other hand, when $2(c_s^m c_t^n \beta_{st})^2 \gg E_m^* - E_n^*$, the covalent term becomes predominant and a "frontier-controlled" interaction results; in such cases the extent of perturbation, that is, electron transfer, is large. Frontier-controlled processes are favored by conditions that are inimical to charge control, that is, factors that minimize the electrostatic term or maximize the charge transfer contribution of equation (45). These include the presence of weakly polar species lacking high charge density but possessing reactive atoms with high base HOMO and acid LUMO densities, species with large orbital radii (although β_{st} increases at small distances, the corresponding increase in the Coulombic term is much greater), bases of low electronegativity (low IP) and acids of high EA, high degrees of overlap and symmetry, and near degeneracy of HOMO and LUMO, that is, $E_m^* - E_n^* \simeq 0$. Charge transfer occurring in frontier-controlled processes leads to a large gain in stability (Figure 4.1*b*).

Acids that favor charge-controlled interactions tend to react with bases that favor the same; an analogous statement holds for acids and bases inclined towards frontier-controlled reactions. Adducts formed across charge-frontier lines lack both the stabilities associated with charge control and frontier control, and are thus very weak.

Since most species (with the exceptions of H^+, He^{2+}, Li^{3+}, etc.) contain both a HOMO and a LUMO (or corresponding atomic orbitals in the case of monatomic species), any substance can be amphoteric, in principle.[13] The relative energies of the HOMOs and LUMOs of two interacting substances must be considered to determine which behaves as an acid, and which as a base, bearing in mind that orbitals close in energy tend to dominate covalent interactions [equation (45)]. The HOMOs and LUMOs of three hypothetical substances X, Y, and Z are compared on an arbitrary energy scale in Figure 4.2 to illustrate the relative nature of acidity and basicity. $(\text{HOMO})_Y$ is much closer to $(\text{LUMO})_X$ in energy than $(\text{HOMO})_X$ is to $(\text{LUMO})_Y$; consequently Y is more likely to behave as a base towards X than X towards Y. On the other hand, the energy difference between $(\text{HOMO})_X$ and $(\text{LUMO})_Z$ is less than the corresponding difference between $(\text{HOMO})_Z$ and $(\text{LUMO})_X$; thus X behaves as a base towards Z. In cases where the energy difference between the HOMO of one species and the

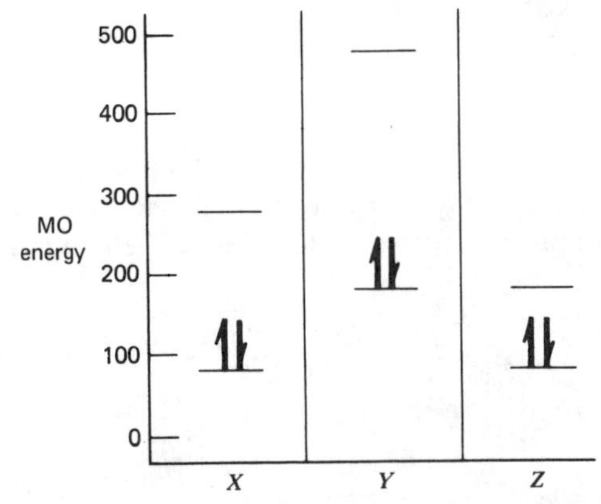

Figure 4.2. Amphoterism as a general acid–base property from a molecular orbital viewpoint.

LUMO of a second approximates the difference between the LUMO of the first and the HOMO of the second, phenomena such as "backbonding" appear.

It is interesting to note that the foregoing discussion makes no assumption about the nature of a HOMO or of a LUMO. This is a departure from the early Lewis formulation, because the requirement, implicit in Lewis' electron dot structures, that both the donor and acceptor orbitals be nonbonding is discarded. A HOMO may be bonding or nonbonding and a LUMO may be nonbonding or antibonding.[13] This assertion offers an explanation of the "electrical strain" postulated by Lewis to explain the ionization of some acid–base adducts and a method of predicting such strain. There are four possible HOMO–LUMO combinations:

1. *Nonbonding HOMO–nonbonding LUMO* No electrical strain is produced and no subsequent dissociation occurs.
2. *Bonding HOMO–nonbonding LUMO* If the loss of electron density by the HOMO upon coordination is sufficient, the base becomes destabilized and simultaneously or subsequently dissociates, since a bond has been broken. Acid–base adduct formation is thus actually an acid displacement reaction [equation (19)].
3. *Nonbonding HOMO–antibonding LUMO* Similarly, a large gain in electron density by the LUMO upon coordination may rupture the acid, since the filling of an antibonding orbital causes destabilization. The reaction then becomes, essentially, a base displacement [equation (18)].
4. *Bonding HOMO–antibonding LUMO* Both acid and base may ionize upon coordination in this case, resulting in a double displacement process [equation (23)].

The Mulliken and Klopman–Hudson equations [equations (42) and (45), respectively] both contain separate electrostatic and charge-transfer terms but differ in that the former describes the total energy of an acid–base system, whereas the latter yields only the change in energy upon adduct formation. The following approximations relate the two approaches[13]:

$$E_0 - (E_A + E_B) \simeq \frac{q_s q_t e^2}{R_{st}\varepsilon} \tag{46}$$

$$\frac{(\beta_{01} - E_{01}S_{01})^2}{E_1 - E_0} \simeq \frac{2(c_s^m c_t^n \beta_{st})^2}{E_m^* - E_n^*} \tag{47}$$

where E_A and E_B are the energies of the isolated A and B species, respectively.

The quantum mechanical formulation of the electronic acid–base theory extends the scope of the concept to all closed shell–closed shell interactions, that is, all processes except oxidation–reduction and free radical reactions.[3,13] In the most general sense, bases are substances initiating reactions using doubly occupied orbitals, while acids employ empty orbitals to participate in reactions. Charge-transfer processes, hydrogen bonding, weak intermolecular forces, chemisorption, and a multitude of other phenomena are included in this expanded treatment, along with all cases covered by Lewis' earlier formulation. A thorough presentation of early and contemporary Lewis theory, with emphasis on the quantum mechanical aspects, is given by Jensen.[13]

4.4 QUANTITATIVE ASPECTS OF THE LEWIS THEORY

A major limitation of the protonic and solvent systems concepts, according to Lewis,[5] was their narrow picture of acid–base strength. The restriction of acids and bases to ionizing solvents, coupled with the leveling effect, was responsible for the definition of acid–base strength as the tendency to liberate solvent ions. The inadequacy of this criterion in nonionizing solvents is clear, and the application of the electronic theory in such cases requires a more general view of acid–base strength. An order of Lewis acid strength can be determined, in principle, by measuring the relative stabilities of adducts formed with a reference Lewis base. Conversely, a relative ordering of Lewis base strengths may be obtained by using a reference Lewis acid. Since all solvents, except inert ones, can act as Lewis acids and/or bases, the Brønsted–Lowry and solvent systems conceptions of strength are included in the more general Lewis approach by regarding the solvent as a reference substance.

Although theoretically sound, this approach fails experimentally because the order of Lewis acid (base) strength is not invariant when the reference substance is changed.[5–8,12–14,17,36,42] Lewis did not believe that a single order of acid or base strength existed, finding wide support for this viewpoint from both proponents and opponents of the electronic theory. The latter cited this deficiency as a reason to retain the protonic concept as the sole valid acid–base theory; Brønsted–Lowry acids and bases follow, in general, invariant orders of strength (barring the leveling effect) because all strengths are related to the reference acid H^+. A comparison of the relative basicities of halide ions towards the proton and the Lewis acid Hg^{2+} indicates that a change of reference substance may not only alter but may sometimes completely invert orders of strength[49]:

$$H^+ \text{affinity: } F^- > Cl^- > Br^- > I^-$$

$$Hg^{2+} \text{affinity: } F^- < Cl^- < Br^- < I^-$$

Thus, for example, HI is a stronger acid than HCl, but HgI^+ is a weaker acid than $HgCl^+$.

Brown and co-workers[50-52] accounted for some of the variations in relative acid–base strengths by noting that steric factors sometimes interfere with adduct formation. For example, trimethylamine is apparently a stronger base than pyridine because $(CH_3)_3N$ displaces C_5H_5N from adducts with BH_3, BF_3, HCl, and HBr. However, pyridine is not displaced from $(CH_3)_3B$; the apparent reversal in base strength arises as a result of steric hindrance between the methyl groups of trimethylboron and trimethylamine. Brown's ideas account for a fraction of the observed orders of strength but do not explain, for instance, the above-mentioned difference between the proton and mercuric ion affinities of the halides.

Luder[7] disagreed with Lewis, contending that a single order of acid or base strength was obtainable simply by determining, from displacement reactions, the electron-pair-donating and accepting tendencies of bases and acids, respectively, without relying on reference acids or bases. Strong acid–strong base adducts are recognized by their high stability; adducts of weak acids and strong bases, or of weak bases and strong acids, possess moderate stability; and weak acid–weak base combinations are notable for the ease with which they undergo acid or base displacement. Luder insisted that the universal order of strength thus determined is subject to only two limitations: the first is the leveling effect, and the second is a concentration effect that seems to vary the order of acid–base strength. For example, hydroxide is a stronger base towards the proton than ammonia is, but Ag^+ seems to prefer NH_3 to OH^-, that is, a precipitate of silver hydroxide dissolves easily in ammoniacal solution. According to Luder, the high concentration of NH_3 relative to OH^- is responsible or this effect, rather than a reversal in the order of basic strength with a change in reference acid. Kolthoff[14] disagreed with this point of view, noting that no AgOH precipitate forms from a solution of silver ion when the hydroxide concentration is very low, but that the silver ion–ammonia complex does form at an equally low ammonia concentration.

ACID–BASE STRENGTH IN CLOSELY RELATED SYSTEMS

One possible approach to quantifying Lewis acidity and basicity is by analogy to the Brønsted concept, that is, the determination of orders of strength in closely related systems or relative to closely related reference substances.[53] Such systems may be chosen so that interferences are negligible or constant, allowing a clear ranking of acids and bases by strength.

Satchell and Satchell[54,55] correlated a large amount of standard free-energy (equilibrium constant) data to obtain generalizations about covalent metal halide acidity. Although coordinate covalent bond strengths are more directly related to enthalpies than to free energies of adduct formation, Satchell and Satchell argued that the latter are of greater practical importance and implicitly assumed that entropy effects in closely related systems are relatively constant.

The factors influencing the acidity of a covalent metal halide MX_n include the following.

1. *Electronic configuration* When $n<4$, MX_n is a strong acid if only one electron pair is required to complete the octet of M, but it is only moderately strong if more than one electron pair is required, since the energy gained upon the addition of one electron pair is small in the latter case, and the accumulated electron density from the coordination of the first electron pair weakens the acid toward further reaction. Within a periodic group the acidity of M drops with increasing atomic number (increasing size), since electron pairs are less strongly attracted by more shielded nuclei. This effect is partially canceled by a simultaneous increase in the ease of hybridization of orbitals of the heavier elements, which tends to increase reactivity.
2. *Substituent effects* Substituents present on M prior to coordination with a base exert both polar and steric effects. The former may be summarized by stating that electron-withdrawing and releasing substituents increase and decrease acidity, respectively. Steric effects become prominent when small metal ions are attached to large, bulky groups.
3. *Effect of base structure* The most important consideration as far as the structure of the base is concerned is the electronic state of the donor. Some donors (notably those with N-, O-, and F-donating atoms) form only single σ bonds upon coordination with acids, in accord with the classical Lewis formulation. Other bases possess π-bond-forming capabilities in addition to the ability to form σ bonds, and still others may act as π acceptors in "back donation." It is of particular importance that the determination of an order of basicity relative to a reference acid involve a set of bases with the same type of bonding capabilities, that is, closely related systems. Otherwise, a change of reference acid, for example, from one with no

backbonding abilities to one with such tendencies, alters and invalidates the order of basicity. If one molecule of acid is likely to react with more than one molecule or ion of base, the effects of successive coordination must be considered. The adduct formed between the acid and the first basic particle cannot be expected to retain properties identical to those of the original acid. On the contrary, the possible steric hindrance of the extra substituent and the accumulated electron density from the first coordination generally weaken the acid towards further reaction. Other base-related factors requiring consideration are the presence of bulky substituent groups on the base (steric factors) and chelation; no meaningful comparison of strength can be made between monodentate and multidentate ligands.

4. *Solvent effects* Since solvents are also Lewis acids and/or bases, solute–solute adduct formation is more likely to be a displacement process, rather than a simple acid–base combination [cf. equations (29)–(34)], and may be influenced by the acid–base strength of the solvent. Solvents of high coordinating power may reverse the apparent relative strengths of, for example, two acids, whose strengths were initially determined relative to a reference solute base in a weakly coordinating solvent, by holding onto the stronger acid via a leveling-type effect, thus lowering its apparent acidity.

Satchell and Satchell predicted that the behavior of a series of related bases towards a covalent metal halide reference acid parallels the behavior of the same bases towards any other covalent metal halide or towards the proton; that is, a linear relationship exists between basicity towards covalent metal halides and towards the proton:[13,54,55]

$$pK_{AB} = a(pK_a)_{HB} + b \tag{48}$$

where K_{AB} is the dissociation constant of A:B under the experimental conditions given, $(K_a)_{HB}$ is the aqueous acid dissociation constant of HB, and a and b are constants that are dependent on the reference acid. Equation (48) is valid when the coordination of only one electron pair takes place and no steric or π-bond effects interfere. Determinations of relative strength also fit equation (48) in multiply coordinated adducts in which the π-bond effects are constant in direction and magnitude.

Other studies have yielded limited orders of Lewis acid or base strength under restricted conditions, that is, relative to a reference solute or solvent. The various techniques employed to obtain these orders include the following: colorimetric measurements and indicator and potentiometric titrations;[56–58] infrared spectrophotometry, in which the vibrational frequency of the reference substance's interacting group decreases linearly with the enthalpy of adduct formation (bond strength);[33,54,59,60] nuclear magnetic resonance, in which the coupling constants between protons and ^{119}Sn, ^{31}P, or ^{19}F in a reference substance increase linearly with enthalpies of adduct formation;[54,61,62] and catalysis experiments.[23]

THE SEARCH FOR A GENERAL APPROACH TO ACID–BASE STRENGTH

Acid–base strength rankings in closely related systems are useful, but they do not provide a general theoretical basis for, or an overall approximation of, relative acid–base strengths. Efforts have been directed towards the attainment of this objective, especially during the last quarter centry.

The influence of charge and size upon acidity and basicity was recognized even before Lewis promulgated a fully developed electronic acid–base theory. Brønsted[10, 63–65] noted that increasing positive charge corresponds to increased acidity and decreased basicity, while increasing negative charge has the opposite effect. Both the charge of the complex and the oxidation state of the central metal atom affect the acidity of aquometal ions.

Cartledge correlated properties of the element with oxidation state, or valence, and size, defining a function he called "ionic potential," Φ, as the valence state of an element Z divided by its ionic radius r:[66]

$$\Phi = \frac{Z}{r} \tag{49}$$

The protonic acid–base theory was predominant at the time of this formulation, so Cartledge formulated that acidic reaction of metal ions as aquometal ion dissociation and that of nonmetallic substances as acid hydrolysis:

$$M^{m+} + nH_2O \rightleftharpoons M(H_2O)_n^{m+} \rightleftharpoons [M(H_2O)_{n-1}(OH)]^{(m-1)+} + H^+ \tag{50}$$

The hydration energy of the first step of equation (50) exceeds the heat of ionization of water for an M^{m+} species of high ionic potential; that is, high ionic potential implies high acidity. Conversely, species of low ionic potential are weakly acidic.

Cartledge proceeded to classify the various valence states of the elements as follows:[67]

If $(\Phi)^{1/2} < 2.2$, the species is basic.

If $(\Phi)^{1/2} > 3.2$, the species is acidic.

If $2.2 \leqslant (\Phi)^{1/2} \leqslant 3.2$, the species is amphoteric.

The concept of hydration is discarded as necessary for the manifestation of acid–base properties in the Lewis theory; consequently Cartledge's classification may be applied directly to the elements.[10] An experimental correlation of charge and size with reactivity confirmed Cartledge's prediction that highly charged, small cations are highly acidic.[68] For example, the strong base sodium sulfide dissolves the sulfides of As(III), Sb(III), and Sn(IV), but not those of Cu(II), Cd(II), and Bi(III). Aqueous ammoniacal solution precipitates trivalent chromium, manganese, and iron, but not divalent manganese, cobalt, or nickel.

Bjerrum[69] sought to compare Brønsted basicity (proton affinity) to Lewis basicity (nucleophilicity) in order to determine whether the former was a sufficient standard by which to correlate ligand strength in transition metal complexes. He compared the relative basicities of the halide ions towards metal ions and found a greater degree of order than had originally been supposed. Contrary to the argument of opponents of the electronic theory, that numerous orders of basicity existed, depending on the reference acid, Bjerrum was able to divide metal ions into two distinct classes: one class exhibits acidic behavior similar to that of the proton; that is, the halide basicities increase in the order $I^- < Br^- < Cl^- < F^-$. Most of the Lewis acids in this group (though not all) have inert gas electronic configurations and are highly electropositive species whose acidity is influenced by their high ionic charge and small ionic radius, for example, Al^{3+} and Fe^{3+}. These metal ions bind primarily electrostatically to small, highly electronegative, nonpolarizable bases, for example, fluoride ion, in preference to bases with the opposite characteristics, for example, iodide ion.

The second class is characterized by behavior previously[49] attributed to mercuric ion; that is, the order of increasing halide basicity is reversed. These Lewis acids are more electronegative than those of the first group; they contain large numbers of outer-shell *d* electrons; and oxidation potential is generally a good measure of relative acidity for the members of this group. These metal ions prefer to bind to large bases of low electronegativity and high polarizability, for example, iodide ion.

Edwards[13,70–73] incorporated the two types of basicity qualitatively noted by Bjerrum into a four-parameter equation:

$$\log\left(\frac{K}{K_0}\right) = \alpha E_n + \beta H \tag{51}$$

where K is a rate or equilibrium constant, K_0 is the same constant in a reference state (water at 25°C), α and β are substrate (Lewis acid)-dependent constants, H is the proton basicity of a substance in terms of the dissociation constant of its conjugate acid, referred to the dissociation constant of the hydronium ion $[(\mathrm{p}K_a)_{\mathrm{H_3O^+}} = -1.74]$,

$$H = \mathrm{p}K_a + 1.74 \tag{52}$$

(H is defined as zero at 25°C in water), and E_n is the nucleophilicity of a substance in terms of its oxidation potential, referred to the following redox couple:

$$\mathrm{H_4O_2} \rightleftharpoons \mathrm{H_4O_2^{2+}} + 2\mathrm{e}^- \qquad E^0 = -2.60 \text{ V} \tag{53}$$

Thus

$$E_n = E^0_{\mathrm{ox}} + 2.60 \tag{54}$$

(E_n is also defined as zero at 25°C in water). Equation (51), the Edwards equation, may be used to correlate rate and equilibrium constants of displacement, complexation, solubility, and other processes with proton basicity and nucleophilicity. Reactions of Lewis bases with high proton basicities are influenced primarily by H, whereas reactions of bases with low proton basicities depend mostly on E_n. Similarly, Lewis acids manifesting proton-like acidity are characterized by high β and low α values; mercuric ion-like Lewis acids possess the opposite characteristics.

Edwards later[13,71,72] amended equation (51) to separate polarizability P and basicity factors in nucleophilicity:

$$E_n = aP + bH \tag{55}$$

The linear relationship between polarizability and nucleophilicity becomes obvious upon rewriting equation (55),

$$\frac{E_n}{H} = a\left(\frac{P}{H}\right) + b \tag{56}$$

and E_n values calculated from polarizabilities agree well with experimental E_n values. Hence the modified Edwards equation is

$$\log\left(\frac{K}{K_0}\right) = AP + BH$$
$$= (\alpha a)P + (\beta + \alpha b)H \tag{57}$$

Reversals in the order of halide affinities for different metal ions were noticed also by Schwarzenbach,[13,74,75] who labeled the proton-like metal ions "class A" acids, and the Hg^{2+}-like metal ions "class B" acids. Ahrland, Chatt, and Davies[12,33,72,76–78] found that the differences between what they called "class (a)" and "class (b)" metal ions hold not only for halide ligands but also for ligands containing donor atoms in periodic groups VA and VIA. They stated that, in general, class (a) metal ions are those that form their most stable complexes with ligands in which the donor atom is the first member of a periodic group (N, O, or F), and class (b) metal ions are those that form their most stable complexes with ligands in which the donor atoms are heavier members of these groups. Therefore the order of metal–ligand complex stability (barring steric hindrance) is

For Class (a) Metal Ions:	For Class (b) Metal Ions:
$N \gg P > As > Sb > Bi$	$N \ll P > As > Sb > Bi$
$O \gg S > Se > Te$	$O \ll S \simeq Se \simeq Te$
$F \gg Cl > Br > I$	$F < Cl < Br < I$

Class (a) character increases with oxidation state, and metal ions belonging to this group prefer to bind to small, highly electronegative ligands (bases). The opposite is true for class (b) metal ions. Thus the stability of adducts with class (b) acids decreases with increasing ligand electronegativity:

$$C \simeq S > I > Br > Cl \simeq N > O > F$$

The order above is almost (but not completely) inverted for adducts involving class (a) metal ions.

Ahrland, Chatt, and Davies postulated backbonding tendencies for the class (b) metal ions, which contain large numbers of *d* electrons, to explain the preference of these metal ions for large bases, which often have weak acceptor tendencies. They also recognized the existence of a borderline class of metal ions possessing properties intermediate between class (a) and class (b).

HARD AND SOFT ACIDS AND BASES

Acid–base strength is not as important as acid–base reactivity on a practical basis. Several approaches towards the quantification of the electronic theory are predicated on the determination of factors affecting reactivity, of which acid–base strength (electron-pair-donating and accepting tendencies) is only one. The first formulation of Lewis acid–base reactivity came from Pearson,[12,72,77,79,80] who linked class (a) and class (b) metal ion acidity to the more general implications of the Edwards equation [equation (57)], that is, the existence of two classes of nucleophiles or bases, one of which includes substances whose reactivities parallel their relative proton basicities, and the second encompassing species whose reactivities parallel their relative polarizabilities. Pearson called the polarizable bases "soft" bases, and the bases following a proton basicity order of reactivity "hard" bases. Soft bases incorporate traits conducive to high polarizability: large size, low oxidation state, high oxidation potential, low electronegativity, and easily distorted or removed valence electrons. Hard bases are small and highly electronegative, have low oxidation potentials, and hold valence electrons tightly.

Class (a) metal ions tend to bind more effectively to hard than to soft bases and are therefore regarded as "hard" acids. Class (b) metal ions ("soft" acids) evidence the opposite tendency. Pearson did not limit hard and soft acids to metal ions but, rather, extended these categories to include all Lewis acids by correlating experimental rate, stability, equilibrium, and other data with hardness and softness.[72,79-81] Any Lewis acid exhibiting a preference for hard bases is a hard acid; such acids are characterized by traits conducive to low polarizability: small size, high positive charge, low oxidation potential, and no easily distorted or removed valence electrons. Soft acids are generally large, have little or no positive charge, and contain easily distorted or removed (often *d* subshell) electrons, all properties contributing to high polarizability.

The division between hard and soft acids and bases is neither sharp nor absolute. Classes of acids and bases with borderline properties do exist; Pearson emphasized that hardness and softness, like amphoterism, are relative qualities. Some soft bases retain high proton affinities, for example, sulfide ion, which is precipitated by (the soft acids) cupric and silver ions, but not by (the hard acids) ferric and aluminum ions. Another manifestation of the ambiguous delineation between hardness and softness is the above-mentioned observation of Ahrland, Chatt, and Davies that the order of adduct stability found with class (a) acids upon increasing donor atom electronegativity (hardness) is not completely inverted for class (b) acids; that is, soft acids do not necessarily form their most stable adducts with the heaviest members of nonmetallic periodic groups VA–VIIA, although hard acids clearly prefer the lightest elements of these groups. Pearson attributed such anomalous behavior to the masking of hardness and softness tendencies by acid–base strength.

The choice of polarizability as a determinant for reactivity is derived from the Edwards equation [equation (57)]. The convenience of the polarizability criterion lies in the fact that the equation is indirectly a basis for the hard and soft acid and base (HSAB) theory, but Pearson[81,82] indicated that properties related to polarization, for example, ionization potential, electronegativity, or oxidation potential, are equally valid reactivity determinants.

Several generalizations can be made from the experimental correlations used by Pearson to classify Lewis acids as hard, soft, or borderline:[72,79–81]

1. Increasing oxidation state increases the hardness and decreases the softness of acids.
2. Increasing electronegativity increases the hardness and decreases the softness of bases.
3. Hardness and softness of donor and acceptor atoms in bases and acids, respectively, are affected by the hardness or softness of ligand groups present prior to adduct formation. This effect is called "symbiosis"[33,77,82,83] and its nature becomes clear when stated simply: soft ligands tend to soften acids and bases, hard substituents harden them. For example, $[Co(NH_3)_5F]^{2+}$ is more stable than $[Co(NH_3)_5I]^{2+}$ because the hard base NH_3 increases the hardness of cobalt, making it more receptive to the fluoride ion than to the iodide ion. On the other hand, substitution of the soft cyanide ligand for ammonia inverts the stability order; $[Co(CN)_5I]^{3-}$ is stable, whereas $[Co(CN)_5F]^{3-}$ does not exist.

Perhaps the most significant corollary of the HSAB theory is the statement that hard acids prefer to react with hard bases, and soft acids with soft bases,[12,13,33,72,73,77,79–81] which serves as the basis for Pearson's interpretation, correlation, and prediction of a great deal of chemistry. These preferences were recognized (although not in terms of acids and bases) during the 19th century by Berzelius, who noted that some metals, for example, Al, Mg, and Ca, always occur in nature as carbonates or oxides, while others, for example, Cu, Pb, and Hg, invariably appear as sulfides.[13,80] Pearson emphasized that this statement is not an absolute principle but a useful rule of thumb to which numerous exceptions exist. Examples of processes explained or predicted by the preference of hard acids for hard bases and soft acids for soft bases include the following:[72,79–82]

1. *Displacement reactions* Acidic (electrophilic) and basic (nucleophilic) substitution tendencies and reaction rates follow a hard–hard and soft–soft preference, for example, the following reaction

$$HI_{(g)} + F^-_{(g)} \rightleftharpoons HF_{(g)} + I^-_{(g)} \qquad \Delta H = -63 \text{ kcal/mole} \qquad (58)$$

is exothermic. Organic reactions are treated analogously by conceptually breaking up the reactants into acidic and basic fragments; for example, methane may be considered as $CH_3^-H^+$ or as $CH_3^+H^-$.[73,80]

2. *Complex stability* AgI_2^- and I_3^- are stable, whereas AgF_2^- and I_2F^- are not. The BF_3CO adduct is unknown, but the BH_3CO adduct is stable; the influence of symbiosis (hard F^- versus soft H^- ligands) on boron may be noted.

3. *Precipitation of insoluble salts* AgF does not precipitate in water because it is a soft–hard combination. AgI, a soft–soft combination, precipitates readily in aqueous solution. AgCl also precipitates, but the chloride ion, not being as soft as bromide or iodide, is easier to dislodge from the silver ion. Therefore AgCl is the only one of the insoluble silver halide salts to dissolve in ammoniacal solution.

4. *Poisoning of catalytic metal surfaces* Metals in the zero oxidation state behave both as soft acids and as soft bases; hence most catalytic poisons are soft acids and bases, for example, CO.

5. *Solvation* Hard solvents prefer hard solutes, and soft solvents soft solutes.[12] Thus many (soft) organic solutes are not very soluble in hard solvents like water. In addition, solvents with hard acid properties, such as water, often level the strength of hard base solutes, but not that of soft base solutes, causing apparent inversion in reactivity. The reaction of methyl iodide, a hard acid–soft base combination, with the hard base fluoride ion to form methyl fluoride and iodide ion serves as an example. When this reaction is carried out in the gas phase, the methyl carbonium ion clearly prefers fluoride to iodide ($\Delta H = -56$ kcal/mole), but in aqueous solution there seems to be a slight preference for iodide ($\Delta H = 2$ kcal/mole), the reason being that the hard acid water competes for fluoride to a much greater extent than for iodide. The energy of fluoride desolvation is large enough to render the overall reaction endothermic.

6. *Oxidation potentials of metals* The oxidation of a metal

$$M_{(s)} \rightleftharpoons M^+_{(aq)} + e^- \tag{59}$$

may be broken up into three steps to obtain a clear picture of the energy consideration involved:

$$M_{(s)} \rightleftharpoons M_{(g)} \qquad \Delta H_{\text{sublim}} \tag{60}$$

$$M_{(g)} \rightleftharpoons M^+_{(g)} + e^- \qquad \text{Ionization potential} \tag{61}$$

$$M^+_{(g)} + e^- \rightleftharpoons M^+_{(aq)} + e^- \qquad \Delta H_{\text{hydrat}} \tag{62}$$

Equations (60) and (61) require the input of energy, and consequently the hydration energy of equation (62) must provide the driving force for the overall reaction. If M^+ is a hard acid, it is better solvated by water, and the oxidation potential of M will be high. If M^+ is a soft acid, it is not well solvated, and the oxidation potential of M will be low.

7. *Invalidity of Pauling's predicted reaction enthalpies*[13] Pauling assumed that the high electronegativity of fluorine, relative to the other halogens, increased the stability of fluoride compounds owing to a high ionic resonance energy (proportional to the square of the difference in electronegativity between fluorine and the other element in a fluoride), and he calculated reaction enthalpies based upon this idea. The magnitudes of Pauling's results agreed with experimental values, but the signs of the enthalpies were wrong. Pearson attributed the discrepancies to Pauling's failure to account for hardness and softness; since the HSAB principle does not assume a single order of reactivity the sign of ΔH may be predicted correctly.
8. *Instability of hard–soft combinations* Equation (58) and the other hard–soft combinations cited above provide examples of such instability. A further example is

$$2CH_2F_{2(g)} \rightleftharpoons CH_{4(g)} + CF_{4(g)} \qquad \Delta H = -26 \text{ kcal/mole} \qquad (63)$$

Pearson[12,33,72,79–81] advanced several possible theoretical interpretations for the hard acid–hard base and soft acid–soft base preferences:

1. *Ionic–covalent bonding* The small size and high positive charge associated with hard acids favor electrostatic bonding; the same is true for highly negatively charged, small (hard) bases. The properties of soft acids are the opposite of the hard acid traits; consequently soft acids favor bonding by orbital overlap (covalency), as do soft bases. Hard–soft combinations are mismatched because very limited stabilization is possible in cases where the bonding tendencies of acid and base differ.
2. *π bonding* Soft acids contain loosely held outer-shell *d* or *p* electrons and are capable of forming π bonds to suitable bases via back donation. Only soft bases have the requisite empty acceptor orbitals available. Hard bases often have more than one pair of electrons available for donation and are able to form doubly dative (one σ and one π) bonds with suitable acids. Hard acids are more likely to

contain a large number of empty, low-energy orbitals than are soft acids. Therefore soft–soft combinations are stabilized by back donation, and hard–hard adducts by doubly dative bonding. Neither reactant possesses electrons for π bonding in hard acid–soft base combinations, and mutual repulsion of π electrons contributes to adduct destabilization in soft acid–hard base coordination complexes.

3. *Electron correlation effects* Extra stabilization is expected for large acceptor atom–large donor atom (soft acid–soft base) combinations owing to the relatively large Van der Waals or London dispersion forces that induce mutual polarization in such adducts. Also, the greater availability of orbitals in larger atoms permits *d–p* orbital hybridization, which increases the overlap and strengthens (lowers the energy of) soft acid–soft base π bonds, simultaneously weakening (raising the energy of) the corresponding antibonding orbitals.
4. *Quantum mechanical perturbation* The Klopman–Hudson treatment of reactivity[13,46–48] may be adapted to the HSAB concept. Circumstances favoring charge-controlled reactions (small size, high charge density, low electron affinity on the part of the acid, low polarizability, high ionization potential on the part of the base, small overlap, and a large difference between base HOMO and acid LUMO energies) are identical with those favoring hard acid–hard base interactions. Conditions favoring frontier-controlled processes (the opposite of those leading to charge control) resemble those favoring soft acid–soft base interactions. Hard–soft combinations are of undefined control and low reactivity.

The early presentation[72] of the HSAB theory was criticized for its lack of quantitativeness. Critics also viewed the invocation of acid–base strength to explain the failure of the concept in certain cases as a handy excuse concealing the total worthlessness of hardness and softness.[72,77,80,81,84] This loophole in the theory also promulgated the misconception that strength and hardness were equivalent.[13] Pearson[12,13,33,73] attempted to respond to these criticisms and clarify the distinction between acid–base strength–weakness and hardness–softness by proposing a four-parameter equation quantifying acid–base reactivity in terms of separate strength and softness factors:

$$\log K = S_A S_B + \sigma_A \sigma_B \tag{64}$$

where K is the equilibrium or rate constant of the reaction of interest. S_A and S_B are parameters that increase in magnitude with increasing acid and base strength, respectively, and σ_A and σ_B are parameters that increase in magnitude with increasing acid and base softness, respectively. Equation (64) emphasizes the independence of strength factors and those properties related to the HSAB concept in reactivity. Pearson's formal inclusion of strength as a determinant in reactivity (instead of as a device to be employed only when HSAB predictions fail) broadens the scope of his ideas only in principle, because displacement and exchange reactions [equations (18), (19), and (23)] are of greater practical importance than the simple adduct formation [equation (9)] upon which equation (64) is based, although Pearson did determine softness numbers for bases as a function of the rate constants of their reactions with *trans*-$Pt(C_5H_5N)_2Cl_2$.[79] Pearson considered equation (64) to be the equivalent of the Edwards equation [equation (51)]. There is some justification for this assumption, since the relationships between base strength S_B and proton basicity H, base softness σ_B and nucleophilicity E_n acid strength S_A and β, and between acid softness σ_A and α all seem to constitute valid analogies.

Pearson[13,33,73] proposed that the proton and methylmercuric ion be taken as hard acid and soft acid empirical standards, respectively, as a practical measure of reactivity against which the relative reactivities of Lewis bases may be determined by measuring the equilibrium constants of

$$CH_3HgH_2O^+ + HB \rightleftharpoons CH_3HgB + H_3O^+ \qquad (65)$$

The advantage of using methylmercuric, rather than mercuric, ion is that the former eliminates complications that would be introduced by the ability of the mercuric ion to form 1:2 adducts. The relative reactivities of Lewis acids could then be determined by exchange equilibria analogous to equation (65), using standard hard and soft bases. Yatsimirskii[85] published such correlations of pK_{CH_3HgB} versus pK_{HB} and of pK_{ACl} versus pK_{AOH}.

The HSAB concept has been attacked on numerous grounds. The arbitrary conceptual breakup of organic reactants into positive and negative fragments is viewed as an "after the fact" convenience that is useful in illustrating mechanisms of known reactions, but without predictive value in the investigation of unknown processes.[12] Jørgensen[83] noted that Pearson's assertion, that softness decreases and hardness increases with increasing oxidation state, is modified by symbiosis (a term coined by Jørgensen), which must be considered in species incapable of independent existence at high oxidation states. For example, manganese exhibits maximum hardness in the +2 state instead of the +7 state because Mn^{7+} does not exist in the free state; in substances such as permanganate the soft oxide ligands exert a symbiotic effect upon the central metal atom.

Myers[78] disputed the validity of polarizability and oxidation state as criteria for determining hardness and softness, pointing out the following examples of inconsistency in the HSAB concept:

1. Although the molar polarization of Cs^+ exceeds that of Ag^+, Hg^{2+}, Tl^{3+}, Pb^{4+}, and Cu^+, Cs^+ is classified as a hard acid, whereas the others are regarded as soft.
2. The molar polarizations of Tl^{3+} and Pb^{4+} are less than those of Tl^+ and Pb^{2+}, respectively, yet the more highly charged species of each element is regarded as softer.
3. The very high molar polarization of Pb^{2+} implies that it should be one of the softest Lewis metal ion acids, but in an aqueous solution of cyanide ion (a soft base) the $Pb(OH)_3^-$ ion is formed, despite the hardness of hydroxide.
4. Pearson associated polarizability with the presence of large numbers of outer-shell *d* electrons, but the molar polarization of K^+, which contains no *d* electrons, exceeds that of Cu^{2+} and of Zn^{2+}, each of which contains the maximum of 10 outer-shell *d* electrons.
5. Pearson's equating the quantitative formulation of HSAB theory [equation (64)] with the Edwards equation implies a linear relationship between α and polarizability that is not realized in practice.

Myers blamed these inconsistencies on the fundamental, incorrect starting assumption of the HSAB concept, that increasing polarizability increases the tendency towards covalent bonding. On the contrary, polarization of a ligand by a metal ion is an effect that localizes electron density, that is, an electrostatic effect. Both ionic and covalent bondings decrease with increasing size and polarization. It is the difference in electronegativity between an acid and a base that is responsible for covalent bonding, and not polarizability; the smaller this difference, the greater the extent of covalent bonding.

Yatsimirskii[85] believed the hard–soft categories to be too general and proposed a more rigorous and detailed six-category approach towards classifying Lewis acids and bases, based on the type of bonding predominant in their adducts:

1. *Acids and bases with purely electrostatic (Coulomb or Madelung) interactions* No charge transfer processes occur in adducts involving members of this class, and therefore they are not strictly Lewis acids and bases. The acids are large cations lacking empty, low-energy orbitals (Cs^+, Rb^+, NH_4^+) and the bases are large anions lacking free electron pairs, for example, BR_4^-.
2. *Acids and bases that are pure σ acceptors and donors, respectively* Only a single coordinate covalent bond is formed in adducts contain-

ing acids such as CH_3^+ and H^+, or bases such as H^-, CH_3^-, and NH_3.

3. *Acids and bases that are predominantly σ acceptors and donors, respectively, but are also inclined towards double dative bonding* In addition to their σ capabilities, these bases (OH^-, F^-, carboxylate anions, oxyanions, and water) also manifest weak π-donating capabilities, and acids of this group (metal ions with closed electronic configurations, e.g., the lighter elements of periodic groups I A–III A) are weak π acceptors in addition to being σ acceptors.
4. *Acids and bases with strong double dative tendencies* The bases are strong σ and π donors, for example, Cl^-, Br^-, and S^{2-}, while the acids are strong σ and π acceptors, for example, Ga^{3+}, Pb^{2+}, and Sn^{2+}.
5. *Acids containing mobile d electrons* These acids are σ acceptors but are also capable of π donation in backbonding. They include the +2 and +3 ions of iron, cobalt, and copper, and other oxidation states of other transition metal ions.
6. *Bases that are σ donors but π acceptors* These bases can also take part in backbonding and include CO, CN^-, CNS^-, and so on.

Hard acids and bases are equivalent to the second and third categories of the Yatsimirskii classification, while groups (4)–(6) include soft acids and bases. Yatsimirskii and others[85,86] also noted that Pearson's classification disregards experimental factors such as solvation. For example, the HSAB principle classifies the proton as a hard acid and the methyl carbonium ion as somewhat softer, but halide affinities of the two acids determined in the gas phase indicate that they are equally hard, the reason for this discrepancy being that the HSAB classification is based, in large part, upon observations of chemical phenomena in aqueous solution. It is therefore maintained[86] that simple hydration theory (and, by extension to other solvents, solvation theory) explains much of what Pearson claimed was caused by hardness and softness, rendering his concept useless.

Drago devised an alternate approach to reactivity that he deemed superior to the HSAB theory. Discussion of Drago's criticism of the HSAB concept requires prior consideration of the alternate model.

THE *E*–*C* EQUATION

Two years after the introduction of the HSAB principle, Drago and Wayland[87] proposed a four-parameter equation correlating enthalpies of

Lewis acid–base adduct formation with electrostatic and covalent interactions:[13,33,87–90]

$$-\Delta H = E_A E_B + C_A C_B \tag{66}$$

where E_A and E_B are parameters that increase with increasing susceptibility of A and B, respectively, to electrostatic interactions, and C_A and C_B are parameters that increase with increasing susceptibility of A and B, respectively, to covalent interactions.

Drago established a set of empirically derived E and C numbers by first obtaining E_B and C_B values for some of the amines, based on the enthalpies of formation of I_2–amine adducts, and utilizing iodine as the reference acid ($E_A = C_A = 1.00$). He then used the amine numbers to obtain E_A and C_A for other acids, and used the latter numbers to obtain E_B and C_B for other bases, and so on. The enthalpies from which E and C were determined were measured in poorly solvating media to avoid the effect of heat of solvation (which, by altering observed enthalpies, would have rendered E and C numbers solvent dependent); enthalpies measured in the gas phase are better suited to the Drago concept but are experimentally impractical. According to Drago, enthalpy measurements are more indicative of the true stability of a Lewis acid–base adduct than are free-energy (equilibrium) measurements because the former are free of the entropy effects that complicate the determination of coordinate covalent bond strength.

It is obvious from equation (66) that adduct stabilities are particularly great in cases where both the acid and base have large susceptibilities to the same kind of interaction; that is, acids of high E_A tend to react best with bases of high E_B, and acids of high C_A prefer to coordinate with bases of high C_B. The following are regarded as advantages of the E–C equation:[13,33,87–91]

1. The set of E–C numbers is internally consistent. Predictions of ΔH of untried acid–base combinations on the basis of E–C numbers have generally agreed with subsequent experiments.

2. The E–C equation explains reversals in acid–base reactivity. For example, diethyl ether is a stronger base towards phenol than its thioether analogue, but towards iodine the sulfide is a stronger base. This is explained by comparing the E and C numbers of both the acids and the bases: $E_{A(\text{phenol})} > E_{A(I_2)}$; $E_{B(\text{ether})} > E_{B(\text{sulfide})}$; $C_{A(I_2)} > C_{A(\text{phenol})}$; and $C_{B(\text{sulfide})} > C_{B(\text{ether})}$. It may be noted that these preferences depend on two different types of interactions.

3. The E–C equation is consistent with the ionic–covalent concept and other explanations of donor–acceptor interactions. The degrees of

ionicity and covalent character in a given adduct predicted by the $E-C$ numbers often agree with estimates of the same quantities by other methods.

4. A comparison of E and C values of a series of acids or bases indicates that increasing E does not necessarily imply decreasing C; that is, electrostatic and covalent interactions are not mutually exclusive. For example, ICl has both a greater E_A and C_A than does I_2 and is consequently a stronger acid toward bases that bind primarily electrostatically, and also toward those bonding essentially covalently.

5. The $E-C$ equation does not distinguish between different varieties of acid–base bonding because the E and C numbers are predicated on the assumption of a single, coordinate covalent σ bond. However, double dative bonding, backbonding, and intermolecular hydrogen bonding effects are recognized by experimental ΔH values that do not agree with predictions of the $E-C$ equation; in such circumstances $-\Delta H_{\text{expt}} > -\Delta H_{EC}$, since the effect of these interactions is to increase adduct stability. Steric hindrance may similarly be recognized by the destabilization it brings to an adduct ($-\Delta H_{\text{expt}} < -\Delta H_{EC}$).

6. $E-C$ numbers do not represent the ground states of the reacting acid and base but are a measure of their properties during interaction; for example, the ability of nonpolar BF_3 to form highly polar adducts is acknowledged by its large E_A value.

7. The $E-C$ equation is the empirical descendant of the Mulliken[13,43–45] and Klopman-Hudson[13,46–48] quantum mechanical formulations of the electronic theory, which also stress the separation of acid–base adduct formation into electrostatic and covalent interactions. Charge-controlled reactions may be associated with acids and bases with large E values, and frontier control with those having large C values.

8. Results of the $E-C$ equation correlate well with spectral changes (infrared vibrational frequencies, NMR coupling constants) observed upon adduct formation.[59,61]

Pearson[92] contended that the $E-C$ equation is equivalent to his quantification of HSAB [equation (64)], and it was also suggested that softness and hardness can be correlated with C and E, respectively, and that acids and bases with large C/Es can be regarded as soft, while those with small C/Es are hard.[93] Drago[94] responded that his equation does not purport to separate strength from the other factors influencing reactivity (as Pearson's does) and that Pearson, by equating equations (64) and (66), implies that hardness–softness cannot be separated from strength, which is a self-contradiction on the part of the HSAB concept that is intimately related to its fundamental invalidity. Furthermore, C/E is a misleading measure of hardness or softness for two reasons.[88-91] First, E–C numbers are based on an arbitrary reference point ($C_A = E_A = 1.00$ for I_2); a different reference compound or set of reference numbers shifts the entire E–C scale and all E/C accordingly. Second, the significance of the magnitudes of the individual E–C numbers is lost when they are combined in a ratio, for example, the hard acid trimethylaluminum ($E_A = 16.9$, $C_A = 1.49$) has approximately the same C/E as does the soft acid trifluoroacetic acid ($E_A = 5.56$, $C_A = 0.509$).

Drago[87-91] believed that the depiction of hardness and softness as mutually exclusive properties is the fundamental deficiency of the HSAB concept; an increase in hardness does not necessarily imply a decrease in softness, and vice versa. The failure of the HSAB principle to explain certain chemical processes and the failure of the above-mentioned arguments for the equivalence of the E–C and HSAB concepts can be traced to this point. Most acids and bases have both hard and soft character, according to Drago. The inability of the Pearson concept to perceive that one acid (or base) may be both harder and softer than another is the weakness that requires the HSAB theory to arbitrarily invoke strength to explain stable hard–soft combinations.

The E–C approach, on the other hand, permits one substance to be both harder (larger E) and softer (larger C) than another. Acids and bases may be classified as hard or soft on the basis of their E–C numbers, but such classification does not exclude the possibility that, for example, a hard acid like trimethylaluminum ($C_A = 1.49$) can form a more stable adduct with a soft base than the soft acid iodine ($C_A = 1.00$). Thus the E–C equation is not, as Pearson suggested, merely quantitative HSAB theory, but provides instead a satisfactory interpretation for many processes for which the hard–soft rules fail.

Drago[90] and Drago and Kabler[91] compared the enthalpies predicted by the E–C equation to those predicted by three equations that he felt might represent quantitative adaptation of the HSAB theory to further illustrate the superiority of his approach over the HSAB concept:

$$-\Delta H = \underbrace{(K - H_A)(K' - H_B)}_{\text{Softness}} + \underbrace{H_A H_B}_{\text{Hardness}} \tag{67}$$

$$-\Delta H = \frac{1}{H_A H_B} + H_A H_B \tag{68}$$

$$-\Delta H = \frac{k}{H_A H_B} + H_A H_B \tag{69}$$

where H_A and H_B are parameters that increase with increasing hardnesses of A and B, respectively, K and K' are constants that transform the hardnesses of A and B, respectively, into softnesses, and k is a best-fit scaling factor. The E–C equation was found to give better agreement with experimentally observed adduct formation enthalpies than any of the three "hardness–softness" equations.

The enthalpies predicted by the E–C equation agree with experiment only in carefully defined systems.[13] Like the Mulliken and Klopman–Hudson formulations, Drago's approach is implicitly based on a small degree of charge transfer and thus fails for strongly interacting systems ($-\Delta H \gg 50$ kcal/mole). The E–C equation also fails when ions are considered as reactants instead of discrete, neutral species. The concept is inapplicable in strongly solvating media, in which the experimental enthalpies are altered by heats of solvation, and is thus useless in displacement and exchange reactions, which, rather than simple adduct formation, comprise the majority of Lewis acid–base processes. Finally, the E–C equation has been criticized because it disguises the relationship between reactivity and periodic elemental properties through the use of empirical numbers based on an arbitrary reference point.[13]

DONOR AND ACCEPTOR NUMBERS

Gutmann and co-workers[13,95–101] proposed that enthalpies of adduct formation between solvent Lewis bases and reference Lewis acids be considered a measure of the chemical properties of the solvent (solvation, solvolysis, solute dissociation), because all solvent reactions depend on coordination. Using antimony pentachloride as a reference acid, Gutmann defined the "donor number" or "donicity," DN, of a basic solvent B as a measure of its strength:

$$\mathrm{DN} = -\Delta H_{B:\mathrm{SbCl}_5} \tag{70}$$

DN includes the total interaction of B with $SbCl_5$ (electrostatic and covalent) and may therefore be regarded as a quantitative measure of base reactivity.

Gutmann obtained experimental $\Delta H_{B:\mathrm{SbCl}_5}$ values from calorimetric measurements of the heats of reaction of $SbCl_5$ and various solvent bases in 1,2-dichloroethane. Gas-phase measurements, although preferable for enthalpy determinations, since there is no solvation to affect ΔH, are generally precluded by associated experimental difficulties. 1,2-dichloroethane is a relatively inert solvent that provided Gutmann with good reaction conditions without having to resort to complex techniques.

The following experimental evidence supports the validity of the DN scale of base reactivity[95–102]:

1. Spectrophotometric determination of the equilibrium constant of the reaction between $SbCl_5$ and basic solvents,

$$\mathrm{K}_{B:\mathrm{SbCl}_5} = \frac{[\mathrm{SbCl}_5][B]}{[B:\mathrm{SbCl}_5]} \tag{71}$$

   in 1,2-dichloroethane yielded $pK_{B:\mathrm{SbCl}_5}$ values paralleling donor numbers; that is, entropy effects are relatively constant for the same reference acid.

2. A linear relationship exists between the donicities, as defined by Gutmann, and enthalpies of reaction of the same basic solvents with other Lewis acids, for example, phenol, iodine, trimethyltin iodide, and antimony tribromide, in inert solvents such as CCl_4 and $ClCH_2CH_2Cl$; that is, donor properties (including relative base strength) are independent of the reference acid. This relationship can be formulated for a reference acid A as

$$-\Delta H_{B:A} = a(\mathrm{DN}) + b \tag{72}$$

   where *a* and *b* are constants dependent on the reference acid.

3. Spectral changes correlate with DN. The NMR chemical shifts of ^{19}F (in CF_3I), ^{23}Na (in $NaClO_4$ or $NaBF_4$), and ^{29}Si (in silanol) in basic solvents vary linearly with donicity, as does the ^{119}Sn—proton coupling constant of $(CH_3)_3SnI$. The variation in O–H bond length determined by infrared spectrophotometric measurements with changing solvent is also DN dependent.

4. Ligand substitution reaction rates are functions of donor number.

5. Conductivity measurements indicate that the degree of ionization of acidic, covalent solutes in basic solvents is proportional to DN. These experiments were conducted in nitrobenzene, a relatively inert (low-DN) solvent with a dielectric constant high enough not to inhibit dissociation, since conductivity measurements detected free ions but not ion pairs.

Gutmann[13,101,104] also proposed a function analogous to DN for solvents with acidic properties. The "acceptor number" AN of a solvent is derived from the ^{31}P-NMR chemical shift of the adduct formed by the solvent and triethylphosphine oxide in 1,2-dichloroethane. This down-field chemical shift (decrease of electron density around the phosphorus nucleus) increases with increasing solvent acidity. Gutmann assigned a reference value of 100 to the chemical shift of the $(C_2H_5)_3PO:SbCl_5$ adduct and established a scale of acceptor numbers by normalizing the chemical shifts of other $(C_2H_5)_3PO$-acceptor complexes to that value. The largest chemical shifts are found for protic acids; somewhat smaller shifts characterize acidic, organic compounds, for example, chloroform and dichloromethane; and aprotic acids have the smallest chemical shifts.

Triethylphosphine oxide offers numerous advantages as a reference base. It is a strong and very soluble base; the coordination site (oxygen) is sufficiently remote from the phosphorus nucleus so that there is little danger of interference with the ^{31}P resonance; the ethyl groups are large enough to shield any alternate coordination sites, but too small to sterically hinder coordination of acidic solvents with oxygen, making for a well-defined coordination site.

Considering solution processes in terms of Lewis acid–base interactions allows a clear understanding and separation of the factors contributing to the dissociation of covalent solutes in solvents. In this respect the Gutmann viewpoint is essentially an extension of Brønsted's view of protic solvents as acids and bases to aprotic media.

Gutmann[13,97–100,103] noted a general failure to distinguish between ionization, that is, heterolytic cleavage of a covalent substance into an ion pair, and dissociation, the separation of the ions of an ion pair or aggregate. These two processes are independent of each other, the former being dependent on the acceptor–donor (Lewis acid–base) properties of the solvent, and the latter on the solvent dielectric constant ε. The distinction becomes clear upon consideration of, for example, the fact that the weak donor properties of sulfuric acid as a solvent preclude a large degree of ionization even for the strong acid $HClO_4$, despite the near equivalence of the dielectric constants of sulfuric acid and water. Thus the total ionizing power of a solvent depends on solvent acid–base (AN–DN) and dielectric contributions.

The overall dissociation of a covalent solute MX in a basic solvent may be represented by two steps:

$$\text{Ionization (DN)}: \quad B + MX \rightleftharpoons B \dashrightarrow \overset{\frown}{M{-}X} \rightleftharpoons BM^+X^- \tag{73}$$

$$\text{Dissociation } (\varepsilon): \quad BM^+X^- \rightleftharpoons BM^+ + X^- \tag{74}$$

where coordination and solvation stabilize the cation. The anion may also be stabilized by solvation if the solvent possesses some acceptor properties.

The critical stage of ionization [equation (73)] involves a shift of electron density of the M–X bond away from M and towards X, simultaneously with a shift in electron density from B to M. Analogous processes occur in an acidic solvent:

$$\text{Ionization (AN)}: \quad MX + A \rightleftharpoons \overset{\frown}{M - X} \dashrightarrow A \rightleftharpoons M^+XA^- \tag{75}$$

$$\text{Dissociation } (\varepsilon): \quad M^+XA^- \rightleftharpoons M^+ + XA^- \tag{76}$$

where coordination and solvation stabilize the anion (and solvation stabilizes the cation if the solvent has donor properties). Again, the shift of electron density of the M–X bond away from M towards X, this time accompanied by a shift in electron density from X to A, is critical to ionization.

The overall dissociation constant K is equal to the product of the ionization constant K_{ion} [equations (73) and (75)] and the dissociation constant K_{diss} [equations (74) and (76)] in both cases:

$$K = K_{ion} K_{diss} \tag{77}$$

Since most common solvents possess dielectric constants of sufficient magnitude so that dissociation of ionic aggregates is not inhibited, the effect of K_{diss} is usually ignored. However, a solvent of very low dielectric constant should not be regarded as a weaker acid (or base) compared to one of high dielectric constant merely on the basis of evidence relating to the degree of solute ionization, for example, conductivity, because no information concerning the actual acid–base (AN–DN) properties may be drawn from such data.

DN and AN are useful guides for selecting the appropriate solvent for a chemical reaction.[13,98–101,104] For example, it would be useless to attempt an acid–base titration in a solvent with stronger donor properties than those of the titrant base. Complex formation tendencies, covalent solute ionization, and solvation may be estimated for different solvents from DN and AN.

The DN–AN concept has been criticized as more limited in scope than the HSAB principle or the E–C equation because it incorrectly assumes single orders of acid and base strength. This is a consequence of the fact that most of the donors employed by Gutmann are hard bases, that is, have oxygen or nitrogen donor atoms.[13] Gutmann[98] conceded that the linear relationship between DN and adduct formation enthalpies relative to other Lewis acids [equation (72)] breaks down for very hard and very soft metal ion acids, especially if the solutes from which they originate are hard acid–hard base or soft acid–soft base combinations. Although equations

(73) and (75) imply that increasing softness, that is, polarizability or ability to shift electron density, is conducive to ionization, in practice this is true only up to a point. Certainly the limitation of donor and acceptor numbers to solvent molecules does not allow quantitative comparisons with acidities and basicities of ions.

On the other hand, Gutmann's concept offers an interpretation of Lewis acid–base phenomena in strongly interacting systems (highly coordinating solvents), in which the $E-C$ model fails. Furthermore, each DN and AN value is based directly on an experimental measurement, while the $E-C$ numbers are derived from an arbitrary reference point and HSAB correlations are, for the most part, merely qualitative.

REFERENCES

1. Meyer, K. H., *Ber.*, **41**, 2568 (1908).
2. Folin, O., and Flanders, F. F., *J. Am. Chem. Soc.*, **34**, 774 (1912).
3. Luder, W. F., and Zuffanti, S., "The Electronic Theory of Acids and Bases," 2nd rev. ed., Dover, New York, 1961.
4. Lewis, G. N., "Valence and the Structure of Atoms and Molecules," Chemical Catalog Co., New York, 1923.
5. Lewis, G. N., *J. Franklin Inst.*, **226**, 293 (1938).
6. Kolthoff, I. M., "Treatise on Analytical Chemistry," Part I, Vol. I, Kolthoff, I. M., Elving, P. J., and Sandell, E. B., Eds., Wiley-Interscience, New York, 1959, Chap. 11.
7. Luder, W. F., *Chem. Rev.*, **27**, 547 (1940).
8. Hall, N. F., *J. Chem. Educ.*, **17**, 124 (1940).
9. Luder, W. F., *ibid.*, **25**, 555 (1948).
10. Vander Werf, C. A., "Acids, Bases, and the Chemistry of the Covalent Bond," Reinhold, New York, 1961.
11. McReynolds, J. P., *J. Chem. Educ.*, **17**, 116 (1940).
12. Day, Jr., M. C., and Selbin, J., "Theoretical Inorganic Chemistry," 2nd ed., Reinhold, New York, 1969.
13. Jensen, W. B., *Chem. Rev.*, **78**, 1 (1978).
14. Kolthoff, I. M., *J. Phys. Chem.*, **48**, 51 (1944).
15. Sidgwick, N. V., "Electronic Theory of Valency," Oxford, New York, 1927.
16. Luder, W. F., *J. Chem. Educ.*, **19**, 24 (1942).
17. Gyenes, I., "Titration in Nonaqueous Media," Van Nostrand, Princeton, New Jersey, 1967.
18. Luder, W. F., McGuire, W. S., and Zuffanti, S., *J. Chem. Educ.*, **20**, 344 (1943).
19. Lewis, G. N., and Seaborg, G. T., *J. Am. Chem. Soc.*, **61**, 1886 (1939).
20. Lewis, G. N., and Seaborg, G. T., *ibid.*, **61**, 1894 (1939).
21. Lewis, G. N., and Seaborg, G. T., *ibid.*, **61**, 1894 (1939).
22. Luder, W. F., and Zuffanti, S., *Chem. Rev.*, **34**, 345 (1944).

23. Hubbard II, R. A., and Luder, W. F., *J. Am. Chem. Soc.*, **73**, 1327 (1951).
24. Walden, P., "Salts, Acids, and Bases," McGraw-Hill, New York, 1929.
25. Wickert, K., *Z. Phys. Chem.*, **178**, 361 (1937).
26. Wickert, K., *Z. Elektrochem.*, **47**, 330 (1941).
27. Cruse, K., *ibid.*, **46**, 571 (1940).
28. Smith, G. B. L., *Chem. Rev.*, **23**, 165 (1938).
29. Gutmann, V., and Lindqvist, I., *Z. Phys. Chem.*, **203**, 250 (1954).
30. Lindqvist, I., *Acta Chem. Scand.*, **9**, 73 (1955).
31. Huston, J. L., *J. Phys. Chem.*, **63**, 383 (1959).
32. Norris, T. H., *ibid.*, **63**, 389 (1959).
33. Huheey, J. E., "Inorganic Chemistry, Principles of Structure and Reactivity," Harper and Row, New York, 1972.
34. Meek, D. W., and Drago, R. S., *J. Am. Chem. Soc.*, **83**, 4322 (1961).
35. Audrieth, L. F., and Moeller, T., *J. Chem. Educ.*, **20**, 219 (1943).
36. Bell, R. P., *Q. Rev. Chem. Soc.*, **1**, 113 (1947).
37. Bjerrum, J., *Naturwissenschaften*, **38**, 461 (1951).
38. Luder, W. F., *J. Chem. Educ.*, **22**, 301 (1945).
39. Bjerrum, J., *Acta Chem. Scand.*, **1**, 528 (1947).
40. De Carvalho Ferreira, R., *J. Chem. Phys.*, **19**, 794 (1951).
41. Luder, W. F., *ibid.*, **20**, 525 (1952).
42. Gillespie, R., "Proton Transfer Reactions," Caldin, E., and Gold, V., Eds., Chapman and Hall, London, 1975, Chap. 1.
43. Mulliken, R. S., *J. Chem. Phys.*, **19**, 514 (1951).
44. Mulliken, R. S., *J. Phys. Chem.*, **56**, 801 (1952).
45. Mulliken, R. S., *J. Am. Chem. Soc.*, **74**, 811 (1952).
46. Hudson, R. F., and Klopman, G., *Tetrahedron Lett.*, **12**, 1103 (1967).
47. Klopman, G., and Hudson, R. F., *Theor. Chim. Acta*, **8**, 165 (1967).
48. Klopman, G., *J. Am. Chem. Soc.*, **90**, 223 (1969).
49. Lingafelter, E. C., *ibid.*, **63**, 1999 (1941).
50. Brown, H. C., Schlesinger, H. I., and Cardon, S. Z., *ibid.*, **64**, 325 (1942).
51. Brown, H. C., and Adams, R. M., *ibid.*, **64**, 2557 (1942).
52. Brown, H. C., *ibid.*, **67**, 374 (1945).
53. Bell, R. P., "The Proton in Chemistry," 2nd ed., Cornell University Press, Ithaca, New York, 1973.
54. Satchell, D. P. N., and Satchell, R. S., *Chem. Rev.*, **69**, 251 (1969).
55. Satchell, D. P. N., and Satchell, R. S., *Q. Rev. Chem. Soc.*, **25**, 171 (1971).
56. Peterson, W. S., Heimerzheim, C. J., and Smith, G. B. L., *J. Am. Chem. Soc.*, **65**, 2403 (1943).
57. Hawke, D. L., and Steigman, J., *Anal. Chem.*, **26**, 1989 (1954).
58. Hawkins, T. R., and Freiser, H., *J. Am. Chem. Soc.*, **78**, 1143 (1956).
59. Joesten, M. D., and Drago, R. S., *ibid.*, **84**, 3817 (1962).
60. Cook, D., *Can. J. Chem.*, **41**, 522 (1963).

61. Bolles, T. F., and Drago, R. S., *J. Am. Chem. Soc.*, **88**, 5730 (1966).
62. Deters, J. F., McCusker, P. A., and Pilger, Jr., R. C., *ibid.*, **90**, 4583 (1968).
63. Hall, N. F., *J. Chem. Educ.*, **7**, 782 (1930).
64. Kolthoff, I. M., and Bruckenstein, S., "Treatise on Analytical Chemistry," Part I, Vol. I, Kolthoff, I. M., Elving, P. J., and Sandell, E. B., Eds., Wiley-Interscience, New York, 1959, Chap. 12.
65. Brønsted, J. N., *J. Phys. Chem.*, **30**, 777 (1926).
66. Cartledge, G. H., *J. Am. Chem. Soc.*, **50**, 2855 (1928).
67. Cartledge, G. H., *ibid.*, **50**, 2863 (1928).
68. Campbell, J. A., *J. Chem. Educ.*, **23**, 525 (1946).
69. Bjerrum, J., *Chem. Rev.*, **46**, 381 (1950).
70. Edwards, J. O., *J. Am. Chem. Soc.*, **76**, 1540 (1954).
71. Edwards, J. O., *ibid.*, **78**, 1819 (1956).
72. Pearson, R. G., *ibid.*, **85**, 3533 (1963).
73. Pearson, R. G., *J. Chem. Educ.*, **45**, 581 (1968).
74. Schwarzenbach, G., *Experientia Suppl.*, **5**, 162 (1956).
75. Schwarzenbach, G., *Advan. Inorg. Chem. Radiochem.*, **3**, 257 (1961).
76. Ahrland, S., Chatt, J., and Davies, N. R., *Q. Rev. Chem. Soc.*, **12**, 265 (1958).
77. Kettle, S. F. A., "Coordination Compounds," Appleton-Century-Crofts, New York, 1969.
78. Myers, R. T., *Inorg. Chem.*, **13**, 2040 (1974).
79. Pearson, R. G., *Chem. Eng. News*, **43**, 90 (1965).
80. Pearson, R. G., *Chem. Br.*, **3**, 103 (1967).
81. Pearson, R. G., *J. Chem. Educ.*, **45**, 643 (1968).
82. Pearson, R. G., and Songstadt, J., *J. Am. Chem. Soc.*, **89**, 1827 (1967).
83. Jørgensen, C. K., *Inorg. Chem.*, **3**, 1201 (1964).
84. Livingstone, S. E., *Q. Rev. Chem. Soc.*, **19**, 386 (1965).
85. Yatsimirskii, K., *Theor. Exp. Chem. (USSR)*, **6**, 376 (1970).
86. Garnovskii, A. D., Osipor, O. A., and Bulgarevich, S. B., *Russ. Chem. Rev.*, **41**, 341 (1972).
87. Drago, R. S., and Wayland, B. B., *J. Am. Chem. Soc.*, **87**, 3571 (1965).
88. Drago, R. S., *Chem. Br.*, **3**, 516 (1967).
89. Drago, R. S., Vogel, G. C., and Needham, T. D., *J. Am. Chem. Soc.*, **93**, 6014 (1971).
90. Drago, R. S., *J. Chem. Educ.*, **51**, 300 (1974).
91. Drago, R. S., and Kabler, R. A., *Inorg. Chem.*, **11**, 3144 (1972).
92. Pearson, R. G., *ibid.*, **11**, 3146 (1972).
93. Yingst, A., and McDaniel, D. H., *ibid.*, **6**, 1067 (1967).
94. Drago, R. S., *ibid.*, **12**, 2212 (1973).
95. Gutmann, V., and Wychera, E., *Inorg. Nucl. Chem. Lett.*, **2**, 257 (1966).
96. Gutmann, V., Steininger, A., and Wychera, E., *Monatsh. Chem.*, **97**, 460 (1966).
97. Gutmann, V., and Mayer, U., *ibid.*, **100**, 2048 (1969).
98. Gutmann, V., *Angew. Chem. Int. Ed. Engl.*, **9**, 843 (1971).

99. Gutmann, V., *Chem. Br.*, **7**, 102 (1971).
100. Gutmann, V., *Fortschr. Chem. Forsch.*, **27**, 59 (1972).
101. Gutmann, V., *Chem. Tech. (Leipzig)*, **7**, 255 (1977).
102. Ehrlich, R. H., Roach, E., and Popov, A. I., *J. Am. Chem. Soc.*, **92**, 4989 (1970).
103. Gutmann, V., *Pure Appl. Chem.*, **27**, 73 (1971).
104. Mayer, U., Gutmann, V., and Gerger, W., *Monatsh. Chem.*, **106**, 1235 (1975).

CHAPTER

5

USANOVICH THEORY AND THE RELATIONSHIP BETWEEN ACID–BASE AND OXIDATION–REDUCTION REACTIONS

5.1 USANOVICH THEORY

Usanovich[1] proposed the most general of the current acid–base theories, which he derived from a consideration of the similarities between the conductance curves of $AsCl_3$, $SbCl_3$, and related substances, and those of the hydrogen acids in ether and in other basic solvents. The fact that crystalline, salt-like products are often obtained from the reaction of both $AsCl_3$-type compounds and hydrogen acids with pyridine offered further evidence that the Brønsted–Lowry theory did not cover all cases of acidity. The nearly simultaneous appearance of Lewis' proposal and language barriers prevented Usanovich from comparing the electronic theory to his own formulation.

The protonic theory is limited, according to Usanovich, by its arbitrary division of cations into two groups:[1,2]

1. "Onium" cations, excluding completely substituted onium ions, which are Brønsted acids, for example, ammonium ion and its primary, secondary, and tertiary alkyl derivatives; hydronium ion; and aquometal ions.
2. Free metal and completely substituted onium cations, which are not Brønsted acids, for example, quaternary alkyl ammonium ions.

This division is derived from the fact that the Brønsted–Lowry concept discards salt formation and neutralization as characteristic acid–base processes in favor of conjugate acid–base formation, for example

$$\underset{\text{Acid}_1}{HCl} + \underset{\text{Base}_2}{C_5H_5N} \rightleftharpoons \underset{\text{Base}_1}{Cl^-} + \underset{\text{Acid}_2}{C_5H_5NH^+} \tag{1}$$

A typical aqueous neutralization, for example, that of aqueous HCl with aqueous NaOH, is represented by the following net ionic equation:

$$H_3O^+ + OH^- \rightleftharpoons 2H_2O \tag{2}$$

There remains a solution of sodium ions and chloride ions, which, if equivalent amounts of acid and base are employed, is commonly regarded as neutral. This is fundamentally inconsistent, according to Usanovich, because the sodium chloride solution contains no acid to compensate for the basicity of chloride ion [equation (1)] and to justify considering the solution as neutral. On the other hand, in neutralization processes involving bases capable of forming onium cations, for example, NH_3 [or pyridine in equation (1)], the cationic neutralization products are the acids that provide the logical compensation for anion basicity:

$$H_3O^+ + Cl^- + NH_3 \rightleftharpoons H_2O + Cl^- + NH_4^+ \tag{3}$$

thus creating a distinction between bases that exist in combination with metal cations and onium ion-forming bases.

Brønsted justified the exclusion of salt formation as a characteristic acid–base process on the grounds that the products of acid–base reactions are not always salts, and that salts may also result from other types of reactions, for example, oxidation–reduction (electron transfer) or other transfers: [2,3]

$$Na + Cl \rightarrow NaCl \tag{4}$$

$$(CH_3)_3N + CH_3I \rightarrow (CH_3)_4NI \tag{5}$$

$$(CH_3)_2NH + CH_3I \rightarrow (CH_3)_3NHI \tag{6}$$

Equations (5) and (6), neither of which is a Brønsted–Lowry acid–base reaction, may be compared to

$$(CH_3)_3N + HI \rightarrow (CH_3)_3NH^+ + I^- \tag{7}$$

which is. Usanovich pointed out that the products of equations (6) and (7) are identical; that is, in the former equation an acid and a base are formed from a base and a substance (methyl iodide) not considered as an acid by Brønsted. This same nonacid reacts with a base in equation (5) to form another nonacid $[(CH_3)_4N^+]$ and a base (I^-). The arbitrary distinction between these related reactions separates completely substituted onium cations from other onium cations in the protonic theory.

The examples cited above distinguish between protonic and aprotic substances, a distinction that Usanovich believed to be artificial because of the obvious similarities present in processes such as equations (2) and (4), or equations (5), (6), and (7). The existence of acid–base phenomena in aprotic solvents, for example, $COCl_2$ and SO_2, served to further convince Usanovich that Brønsted and Lowry had only partially generalized acid–base theory by refuting hydroxide ion as the only carrier of basic properties, while retaining the proton as the sole manifester of acidic traits (although Usanovich did not deny the special nature of the proton, as will be seen shortly).

According to Usanovich, any process leading to salt formation is an acid–base reaction, be it of the Brønsted–Lowry type or not. The salt-formation criterion allows consideration of the processes originally prompting Usanovich's proposal as acid–base reactions:

$$AsCl_3 + C_5H_5N \rightleftharpoons C_5H_5NAsCl_2^+ + Cl^- \tag{8}$$

The generality of such a rule required a reevaluation and extension of the then prevalent acid–base definitions. Usanovich therefore defined an acid as a substance capable of splitting off any cation or combining with any anion, and a base as a substance capable of splitting off any anion or combining with any cation.[1–8]

The Usanovich definitions expand the acid category to include a wide range of substances:

1. *Hydrogen acids* Split off protons or, at times, combine with anions; for example, HF combines with fluoride ion.
2. *Metallic compounds* Split off metal cations.
3. *Onium compounds* Split off carbonium, ammonium, and so on, ions of varying degrees of substitution.
4. *Acid "anhydrides"* Substances such as SO_3 and CO_2, which combine with anions or negative parts of molecules.
5. *Nonmetallic substances* For example, $AsCl_3$, which splits off $AsCl_2^+$ and/or $AsCl^{2+}$.

Bases include the same substances regarded as such in the protonic theory:

1. *Halides, oxides, sulfates, and so on* Each splits off its respective anion.
2. *Onium-forming bases* For example, ammonia and pyridine, which combine with protons.

Thus the following reactions are all considered acid–base processes:

$$\underset{\text{Base}_2}{Na_2O} + \underset{\text{Acid}_1}{SO_3} \rightleftharpoons \underset{\text{Acid}_2}{2Na^+} + \underset{\text{Base}_1}{SO_4^{2-}} \tag{9}$$

$$3KCN + Fe(CN)_3 \rightleftharpoons 3K^+ + Fe(CN)_6^{3-} \tag{10}$$

$$3(NH_4)_2S + Sb_2S_5 \rightleftharpoons 6NH_4^+ + 2SbS_4^{3-} \tag{11}$$

It may be noted that the salts formed in such reactions are themselves combinations of acidic (cationic) and basic (anionic) species.

Amphoterism is acknowledged as a general property of polar and ionic species by the Usanovich definitions, since, for example, $AsCl_3$ can be regarded as an acid because it splits off $AsCl_2^+$, and as a base because it also splits off chloride ion. However, the recognition of amphoterism does not preclude the predominance of either acidic or basic properties in specific cases. Thus an aqueous NaCl solution is neutral because it contains an acid and a base, both of which are too weak for the properties of one to predominate, while in aqueous HCl, NH_4Cl, or CH_3COONa solutions one of the constituent ions affects acidity and basicity to a greater extent than the other.

The Usanovich theory is the only acid–base concept to explicitly include oxidation–reduction processes as a subclass of acid–base reactions[1–8] if the acid–base definitions are reworded to define acids as substances splitting off electropositive particles or combining with electronegative particles, and bases as substances capable of exactly the opposite behavior. The electron can be regarded as a base according to these versions of the definitions. Therefore, in equation (4), the base sodium donates an electron to the acid chlorine to produce the acid sodium ion and the base chloride ion, the constituents of the salt sodium chloride. Usanovich thus related oxidizing agents to acids, and reducing agents to bases.

Acidity and basicity (excluding that derived from oxidation–reduction) are attributed to coordinate unsaturation in the Usanovich theory. Acids contain coordinately unsaturated electropositive atoms and tend to combine with negative species; bases contain coordinately unsaturated electronegative atoms and tend to combine with positive species. Acid–base properties are therefore regarded as the products of essentially electrostatic forces. Most substances incorporate both types of coordinately unsaturated atoms, that is, a restatement of the generality of amphoterism. The predominance of acidity or basicity in a given species depends on the relative valences of its most electropositive and electronegative atoms. If the valence of the former exceeds that of the latter, acidic properties predominate; if, on the other hand, the electronegative atom is of a higher valence, basic tendencies predominate. The weaker set of properties is not destroyed in either case, but may manifest itself under appropriate circumstances. Only when the predominant atom is coordinately saturated does a substance possess solely acidic or basic properties. Carbon dioxide serves to illustrate this aspect of the Usanovich theory:

$$CO_2 \xrightarrow{O^{2-}} CO_3^{2-} \xrightarrow{O^{2-}} CO_4{}^{4-} \qquad (12)$$

CO_2 contains a tetravalent, coordinately unsaturated, electropositive carbon atom that imparts acid character to the compound (since oxygen is divalent). The addition of a basic oxide ion converts CO_2 to carbonate ion, a basic anion that nevertheless retains some weak acid properties because the carbon atom is still coordinately unsaturated. Further oxide ion addition converts carbonate to the orthocarbonic acid anion, $CO_4{}^{4-}$, which no longer possesses any acidic tendencies; the carbon atom is now coordinately saturated, and the ion's high negative charge renders it a strong base. Schemes similar to equation (12) may be written for SO_3 (hexavalent sulfur) and SiO_2 (tetravalent silicon), among other substances, and are reminiscent of earlier derivations of ammonoacids and "acids of a system" from hypothetical orthoacids in the early days of the solvent systems theory. Chloroplatinic acid, H_2PtCl_6, is an example of a substance in which coordinate saturation excludes basic properties entirely.

Further confirmation of the influence of valence on acidity and basicity may be obtained by comparing oxides and sulfides to halides; the divalence of the former leads to the expectation that, for a given cation, the oxide or sulfide is more basic than the corresponding halides. This prediction is borne out when the basicities of Na_2O and Na_2S are compared to those of the sodium halides. Similar quantitative comparisons may be made among cations of varying valence in combination with the same anion, yielding, for example, the not unexpected confirmation of the prediction that $FeCl_3$ and $AlCl_3$ are more acidic than NaCl. The Usanovich concept thus links acidity and basicity to periodic properties.

The sole exception to the influence of valence, according to Usanovich, is the proton, which indeed possesses special properties; however, he erroneously attributed these properties to the extraordinary small size and high mobility of the proton, while they are actually a consequence of the Grotthus mechanism, to be discussed later. The proton is thus endowed with a far greater degree of acidity than other univalent cations, which accounts for its historical importance in acid–base theory. Therefore HCl is predominantly acidic instead of amphoteric, as would be predicted by comparative valence, and water and ammonia possess amphoteric properties as compared to the strongly basic Na_2O and Na_3N. The special properties of the proton are also reflected in the rapid rates of most protonic acid–base reactions relative to others; for example, $(CH_3)_3NH^+$ can, in principle, be either a proton or a CH_3^+ donor, but it invariably donates the former in acid–base reactions.

Usanovich agreed with Brønsted that quantitative comparisons of acid–base strength should be based on the inverse relationship between conjugate acids and bases. Since each Brønsted acid has only one conjugate base, ordering of strength in the protonic theory is relatively simple, but in the Usanovich theory each acid and base may be related to a multitude of conjugates, complicating the choice of a reference scale.

Usanovich utilized oxidation–reduction to obtain an order of elemental acid–base strength. Electron donors (reducing agents) are bases whose relative strengths vary with their ionization potentials[1,2]; the lower the ionization potential, the stronger the base. Their cationic acid conjugates vary in strength inversely with base strength; that is, the higher the ionization potential of the base, the stronger the conjugate acid. Thus potassium is a strong base and potassium ion a weak conjugate acid because

$$K \rightleftharpoons K^+ + e^- \tag{13}$$

requires an ionization potential of 4.3 eV, while chlorine is a very weak base because it has an ionization potential of 13.0 eV. On the other hand, chlorine exhibits strong acid properties, and its conjugate base, chloride ion, is very weak:

$$Cl^- \rightleftharpoons Cl + e^- \tag{14}$$

These considerations may be extended to compounds. The use of ionization potential to determine relative acid and base strengths also clearly indicates the greater acid strength of multivalent elemental cations (involving the sum of several ionization potentials) relative to univalent cations. Solvolysis of salts is also explained by a consideration of the relative strengths of the acidic (cationic) and basic (anionic) species present in a salt; for example, $FeCl_3$ hydrolyzes in an acidic manner, and Na_2S in a basic manner.

Luder and Zuffanti[7] objected to the importance of salt formation and the emphasis on the association of acidity and basicity with cations and anions, respectively, in the Usanovich concept. They also contended that there is a lack of correlation of the degree of coordinate unsaturation, the determinant of acid–base properties, with the Usanovich acid–base definitions, and that redox reactions, although related to acid–base processes, are not part of acid–base chemistry (this point will be discussed in more detail later). The Usanovich theory has also been criticized for its generality.

It has similarly been claimed that the Usanovich concept is often too easily dismissed precisely because of its generality, and because of the same traditionalism among scientists who initially opposed the protonic and electronic theories. Huheey[6] proposed that all acid–base concepts are reconcilable because they invariably define acids as donors of positive species or acceptors of negative species, and bases as donors of negative species or acceptors of positive species. Consequently acidity may be regarded as a positive character of a substance, decreasing by reaction with a base, and basicity may be regarded as a negative character, decreasing by reaction with an acid.

5.2 THE RELATIONSHIP BETWEEN OXIDATION–REDUCTIONAND ACID–BASE CONCEPTS

Although the Usanovich theory is the only one to explicitly stress the inclusion of redox phenomena in acid–base chemistry, the position of oxidation–reduction in relation to other acid–base concepts has been the subject of extensive consideration, often as a direct consequence of comparison with the Usanovich concept. However, this relationship was noted even before Usanovich made his proposal.

BRØNSTED–LOWRY THEORY

Most presentations of the protonic theory ignore the fact that Brønsted recognized the electron as the only particle exhibiting behavior comparable to that of the proton, and oxidation–reduction reactions (electron transfer) as the only analogues of acid–base processes (proton transfer):[9–14]

$$A \rightleftharpoons B + H^+ \quad (15)$$

$$R \rightleftharpoons Ox + e^- \quad (16)$$

where A is an acid, B a base, R a reduced form of a species, and Ox an oxidized form of the same species. This relationship arose in part from the belief (at the time of Brønsted's proposal) that the proton and electron retained unique positions as the fundamental, subatomic constituents of matter. Thus the postulation of a parallel between protonic and electronic reactions is not coincidental; Brønsted[9] referred to the proton as a "positive electron."

There is more to the analogy between acid–base and oxidation–reduction processes than the formal similarity of equations (15) and (16):

1. Neither the proton nor the electron is capable of independent existence in solution.[6,15,16]

2. An acid cannot donate a proton in the absence of a base; conversely, a base cannot accept a proton unless an acid is present to furnish one. Similarly, a reducing agent cannot donate electrons without the presence of an electron acceptor (oxidizing agent), and an oxidizing agent cannot accept electrons without an electron donor (reducing agent). Consequently oxidation–reduction reactions, like protolyses, are double conjugate pair equilibria[7,15–17]:

$$A_1 + B_2 \rightleftharpoons B_1 + A_2 \tag{17}$$

$$Ox_1 + R_2 \rightleftharpoons R_1 + Ox_2 \tag{18}$$

3. Powerful oxidizing and reducing agents exhibit phenomena in solvents much like the leveling effect observed by Hantsch for strong acids and bases,[17] for example, in aqueous solution

$$Cl_2 + 3H_2O \rightarrow 2H_3O^+ + Cl^- + OCl^- \tag{19}$$

$$2Na + 2H_2O \rightarrow 2Na^+ + 2OH^- + H_2 \tag{20}$$

The distinction between acid–base and oxidation–reduction processes is blurred in equations (19) and (20), since both occur simultaneously. The oxidizing agent Cl_2 increases the solvent lyonium ion concentration, as any acid does, and the reducing agent Na increases the lyate ion concentration, as any base would. Lewis[18] noted that no rapid oxidizing agent stronger than O_2 nor rapid reducing agent stronger than H_2 may exist in water. Lewis' emphasis on the word "rapid" qualifies the thermodynamics of such processes with the fact that many strong oxidizing and reducing agents act upon water so slowly that redox equilibrium, unlike acid–base equilibrium, is rarely achieved in aqueous solution (witness the slow decomposition of aqueous permanganate).

4. There is evidence that oxidation–reduction reactions occur via stepwise electron transfer in cases involving the movement of more than one electron, analogously to the successive removal of protons from polyprotic acids (or, in certain Lewis acid–base reactions, to the formation of successive 1:1, 1:2, etc., adducts).[17,19–21] Most of the evidence supporting this view is mechanistic, for example, the reduction of hydrogen peroxide by ferrous ion,

$$Fe^{2+} + H_2O_2 \rightarrow Fe^{3+} + OH^- + OH \tag{21}$$

$$Fe^{2+} + OH \rightarrow Fe^{3+} + OH^- \tag{22}$$

or the oxidation of stannous ion by ferric ion (in HCl solution),

$$Fe^{3+} + SnCl_4^{2-} \rightarrow Fe^{2+} + SnCl_4^- \tag{23}$$

$$Fe^{3+} + SnCl_4^- \rightarrow Fe^{2+} + SnCl_4 \tag{24}$$

The observation of a Sn^{3+} intermediate in the titration of $SnCl_2$ with potassium dichromate in aqueous HCl has been reported.[22] It is also believed that diatomic chlorine, bromine, and iodine are reduced through the stepwise intermediates Cl_2^-, Br_2^-, and I_2^-, respectively.

5. Depending on the circumstances, the same species may serve either as an oxidizing or as a reducing agent. Aluminum reduces ferrous ion to iron metal, whereas dichromate oxidizes Fe^{2+} to Fe^{3+}. Luder and Zuffanti[7] regarded such behavior as analogous to acid–base amphoterism and utilized the same term to describe it.

6. Redox indicators operate on the basis of equilibria similar to those of acid–base indicators.[17]

Another manifestation of the recognition of an analogous relationship between acid–base and oxidation–reduction reactions is the suggestion that relative acidity and basicity be defined in terms of acidity and basicity potentials, just as redox strength is defined in terms of oxidation–reduction potentials, instead of as a series of acidic and basic dissociation constants.[9,12,13,15,23] The acidity potential E_{ac} was defined by Brønsted[9,23] as the work expended on proton transfer from an electrically neutral system to an arbitrary standard state. The basicity potential E_{bas} is the negative of E_{ac}:

$$E_{\text{ac}} = \frac{RT}{F} \ln a_H = \frac{RT}{F}\left[\ln K_A + \ln\left(\frac{C_A}{C_B}\right)\right] \tag{25}$$

$$E_{\text{bas}} = -E_{\text{ac}} = \frac{RT}{F}\left[\ln K_B + \ln\left(\frac{C_B}{C_A}\right)\right] \tag{26}$$

where K_A and K_B are the Brønsted acidity and basicity constants, respectively, and C_A and C_B are the conjugate acid and base concentrations, respectively, of equation (15) (see Chapter 2).

Schwarzenbach[15] also stressed the acid–base oxidation–reduction analogy, defining a "normal" acidity potential E^0_{ac} (analogous to standard redox potentials) as the value of E_{ac} at equal activities of A and B (the use of activity, as opposed to concentration, has no bearing upon the present discussion). The equation for the potential of an acid–base conjugate pair consequently follows the same form as the equation for determining the potential of a redox couple:

$$\mathrm{E}_{ac} = \mathrm{E}^0_{ac} + \frac{\mathrm{RT}}{\mathrm{F}} \ln\left(\frac{a_{\mathrm{A}}}{a_{\mathrm{B}}}\right) \tag{27}$$

$$\mathrm{E}_{ox} = \mathrm{E}^0_{ox} + \frac{\mathrm{RT}}{\mathrm{nF}} \ln\left(\frac{a_{ox}}{a_{\mathrm{R}}}\right) \tag{28}$$

E^0_{ac} of a solute acid does not depend on the acidity or basicity of the solvent, but is influenced by the solvent dielectric constant, since it represents the electrical work involved in separating H^+ from an electrically neutral environment.

Arrangement of E^0_{ac} values for different acids in decreasing order yields a ranking according to strength of acids and their conjugate bases similar to the electromotive series in oxidation–reduction. A high acidity potential is associated with substances containing loosely bound, easily transferred protons, that is, strong acids. Weak acids have low E^0_{ac} values associated with tightly bound protons.[13] When two conjugate pairs of different acidity [as measured by E_{ac} in equation (27)] are brought together, proton transfer occurs from the system of higher acidity potential to that of lower E_{ac}, lowering the former and raising the latter until the acidities of both systems are equal, that is, equilibrium is achieved. For example, the dissociation of an acid HX in water involves the acidity potentials of the HX–X^- and the

$H_3O^+–H_2O$ conjugate pairs:

$$E_{HX} = E^0_{HX} + \frac{RT}{F}\ln\left(\frac{a_{HX}}{a_{X^-}}\right) \tag{29}$$

$$E_{H_3O^+} = E^0_{H_3O^+} + \frac{RT}{F}\ln\left(\frac{a_{H_3O^+}}{a_{H_2O}}\right) \tag{30}$$

Prior to dissociation $E_{HX} > E_{H_3O^+}$, but as equilibrium is approached, E_{HX} drops and $E_{H_3O^+}$ rises until, at equilibrium, they are equal and

$$\exp\left[\left(E^0_{HX} - E^0_{H_3O^+}\right)(F/RT)\right] = \frac{a_{H_3O^+}a_{X^-}}{a_{H_2O}a_{HX}} \propto K_a \tag{31}$$

where K_a is the protolysis constant of HX in water, that is, the classical aqueous dissociation constant.

Brønsted[23] pointed out that there is neither advantage nor disadvantage in arranging acid–base strengths as an emf series instead of as a series of equilibrium constants, but the very suggestion illustrates and emphasizes the fact that a close parallel between acid–base and oxidation–reduction processes was recognized well before the appearance of the Usanovich theory.

The relationship between protonic acid–base and redox chemistry is not limited to parallel formalisms because the two sometimes overlap, for example, in equations (19) and (20). The increase in aquometal ion acidity with increasing central metal oxidation number observed by Brønsted[24] is but one example of the increased ability of proton-containing ligands to act as proton donors when coordinated to highly oxidized metal ions.[25] The reactions between aqueous acid solutions or acidic melts and metals are simultaneously acid–base neutralization and oxidation–reduction processes[14,26,27]:

$$2H_3O^+ + Mg_{(s)} \rightleftharpoons Mg^{2+} + H_{2(g)} + 2H_2O \tag{32}$$

$$2NH_4NO_{3(l)} + Cu_{(s)} \rightleftharpoons Cu(NO_3)_{2(l)} + H_{2(g)} + 2NH_{3(g)} \tag{33}$$

Comparing the lower (basic or reducing) oxides of various elements to their higher (acidic or oxidizing) oxide counterparts, for example, Bi_2O_3 to Bi_2O_5 or FeO to Fe_2O_3, Bancroft[28] noted that the emf of the former are raised in alkaline media, while the emf of the latter are higher in acid media. This observation implies that reduction is enhanced in alkaline solution and oxidation is enhanced in acid solution, and this is supported by the variation in the standard potentials of redox couples in aqueous solutions of changing acidity (Table 5.1). Increasing acidity increases the oxidizing ability of a species, while increasing basicity increases reducing power (which becomes evident by reversing the equations of Table 5.1 and noting the sign change in E^0).

TABLE 5.1 Variation in Standard Potentials of Selected Redox Couples With Changing Solution Acidity[a]

	E^0(V)
$Cr_2O_7^{2-} + 14H^+ + 6e^- \rightleftharpoons 2Cr^{3+} + 7H_2O$	1.33
$HCrO_4^- + 7H^+ + 3e^- \rightleftharpoons Cr^{3+} + 4H_2O$	1.20
$CrO_4^{2-} + 4H_2O + 3e^- \rightleftharpoons Cr(OH)_3 + 5OH^-$	−0.12
$MnO_4^- + 4H^+ + 3e^- \rightleftharpoons MnO_2 + 2H_2O$	1.68
$MnO_4^- + 2H_2O + 3e^- \rightleftharpoons MNO_2 + 4OH^-$	0.59
$H_3BO_3 + 3H^+ + 3e^- \rightleftharpoons B + 3H_2O$	−0.73
$H_2BO_3^- + H_2O + 3e^- \rightleftharpoons B + 4OH^-$	−2.5
$H_3PO_4 + 2H^+ + 2e^- \rightleftharpoons H_3PO_3 + H_2O$	−0.28
$PO_4^{3-} + 2H_2O + 2e^- \rightleftharpoons HPO_3^{2-} + 3OH^-$	−1.05
$ReO_4^- + 8H^+ + 7e^- \rightleftharpoons Re + 4H_2O$	0.37
$ReO_4^- + 4H_2O + 7e^- \rightleftharpoons Re + 8OH^-$	−0.81
$Cr^{6+} + 3e^- \rightleftharpoons Cr^{3+}$ (1 F H_2SO_4)	0.69
$Cr^{6+} + 3e^- \rightleftharpoons Cr^{3+}$ (0.01 F NaOH)	0.46
$Fe(CN)_6^{3-} + e^- \rightleftharpoons Fe(CN)_6^{4-}$ (2 F H_2SO_4)	1.10
$Fe(CN)_6^{3-} + e^- \rightleftharpoons Fe(CN)_6^{4-}$ (1 F NaOH)	−0.12

[a]Source: Weast, R.C., Ed., *Handbook of Chemistry and Physics*, 51st ed., The Chemical Rubber Co., Cleveland, OH, 1970.

The individual nature of an acid or base also affects redox potentials, as does the level of acidity or basicity in a solution. For example, the standard reduction potential of the ferric–ferrous ion couple (0.77 V) decreases in acidic solutions because the ferric ion complexes with acid anions to a greater extent than does the ferrous ion. However, the difference in the extent of complexation with both ions varies with the nature of the anion, and consequently the drop in potential is not the same in the presence of different acids (in 1 F aqueous solutions of each of the following acids, the Fe(III)–Fe(II) standard potential is $HClO_4$, 0.73 V; HCl, 0.70 V; H_2SO_4, 0.61 V; and H_3PO_4, 0.44 V). Such effects are the basis for the concept of "formal potential," a quasistandard oxidation–reduction potential defined for a given redox couple under specific experimental conditions, for example, in the presence of a given concentration of a certain acid.[29]

The acid–base and oxidation–reduction concepts are linked, in Brønsted's view, by the equation for the oxidation of hydrogen:[9,14]

$$\tfrac{1}{2}H_2 \rightleftharpoons H^+ + e^- \tag{34}$$

A comparison of equation (34) to equations (15) and (16) indicates the special positions that Brønsted accorded to hydrogen, as simultaneously the simplest acid and reducing agent, to the electron, as the simplest base, and to the proton, as the simplest oxidizing agent. Application of the mass-action law to equation (34) at a hydrogen gas pressure equal to 1 atm yields an equilibrium constant K_N:[9]

$$K_N = a_H a_e \tag{35}$$

where a_e is a measure of reducing capacity (just as a_H is a measure of acidity). The tendency of protons and electrons to combine imparts a very small magnitude to K_N. The hydrogen ion activity in aqueous solutions has a finite lower limiting value set by the autoprotolysis constant of water, K_W. Consequently, equation (35) suggests that the concentration of electrons in aqueous solution is too small to be measurable, at least over the major portion of the range of acidity possible in that medium. Brønsted believed that equation (34) is shifted rightward in more basic solvents, in which measurable concentrations of "free" (in the sense of merely being solvated) electrons might exist. Another method for obtaining measurable concentrations of electrons is to use a reducing agent stronger than hydrogen, for example, sodium, and in this connection Brønsted cited Kraus' experiments with metals dissolved in liquid ammonia and in amines.[5,6,30–33]

Kraus observed that the dissolution of metals in liquid ammonia produces blue solutions at low metal concentrations. These solutions turn bronze with increasing concentrations, exhibiting definite metallic luster, and eventually separate out from the bulk of the solvent into a less dense, bronze phase at very high concentrations. The dilute solutions are highly conducting, and the conductivity of the bronze phase approaches that of pure metals. The solutions are metastable, decomposing to metal amide solutions with time. Kraus found that the positive conducting species in each solution is the solute metal cation, but he could not find chemical evidence for a complementary conducting anion except that the negative conducting species in each solution is identical, regardless of the metal.

Kraus postulated the electron to be the anionic species in the solutions of metals in liquid ammonia:

$$M + (x+ny)NH_3 \rightarrow [M(NH_3)_x]^{n+} + n[e(NH_3)_y]^- \tag{36}$$

The presence of the solvated electron explains the high conductivity of the dilute metal solutions and the almost metal-like conductivity and properties, for example, luster, at high concentrations, at which the degree of solvation decreases and the bronze phase is formed. This bronze phase is believed to be a mixture of metal ions, ammonia molecules, and a sea of movable free electrons. Electron paramagnetic resonance experiments have confirmed the paramagnetism of these systems and the presence of electrons in an environment close to that of the free electron.[6] An alternate explanation of the high conductivities of these solutions involves a quantum mechanical tunneling effect in which the wave nature of the electron allows it to tunnel through the energy barriers to conduction. The metastability of these solutions is due to the leveling of electrons to amide ions:

$$[e(NH_3)_y]^- \rightarrow NH_2^- + \tfrac{1}{2}H_2 + (y-1)NH_3 \tag{37}$$

Similar phenomena are observed in other basic solvents, for example, ethers,[6] and the resultant solutions are both very basic and very reducing.

More recently the solvated electron has been generated in aqueous solution by several methods, including irradiation with ultraviolet light or ^{60}Co gamma rays, dissolution of amalgamated sodium or trivalent uranium, and from atomic hydrogen in strongly basic solution.[34] The acidity of water, relative to ammonia, imparts a high reactivity and a very short lifetime ($\simeq 10^{-3}$ sec) to $e^-_{(aq)}$.[6,34] The solvated electron reacts with hydrogen acids in water:[35]

$$e^-_{(aq)} + HA \rightleftharpoons HA^- \rightleftharpoons H + A^- \tag{38}$$

and, as with other bases, the extent of protolysis increases with increasing HA dissociation constant, but an unequivocal determination of whether the order of Brønsted acid strength relative to $e^-_{(aq)}$ is identical to that relative to other bases is complicated by the existence of different modes of HA decomposition,[36] for example,

$$e^-_{(aq)} + CH_3COOH \rightleftharpoons CH_3COOH^- \begin{array}{l} \rightleftharpoons H + CH_3COO^- \\ \rightleftharpoons CH_3C{=}O + OH^- \end{array} \tag{39}$$

Hydroxide ion is the only base strong enough to compete for protons with the solvated electron in aqueous solution:[36,37]

$$H + OH^- \rightleftharpoons e^-_{(aq)} + H_2O \qquad pK = 9.6 - 9.7 \tag{40}$$

Equation (40) may be separated into

$$H \rightleftharpoons H^+ + e^- \tag{41}$$

$$H^+ + OH^- \rightleftharpoons H_2O \tag{42}$$

Equation (42) is the reverse of the autoionization of water, and is characterized by a $p(K_W^{-1})$ of -14, indicating that equation (41), which is identical to equation (34), has a pK_N of approximately 24, supporting Brønsted's hypothesis [equation (35)] that, because of the lower limit placed on a_H in water, the concentration of $e^-_{(aq)}$ is infinitesimally small under ordinary conditions.

Huheey[6] considered the proton to be the "ultimate acid," and the electron the "ultimate base," in agreement with Brønsted's view of their fundamental nature, although this viewpoint is not to be construed as excluding aprotic substances from the category of acids.

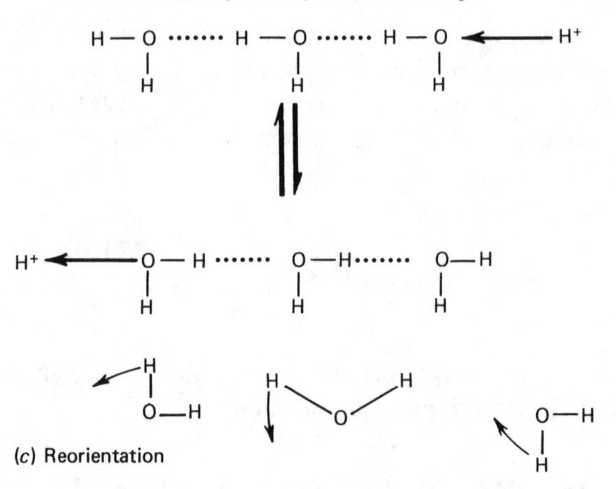

Figure 5.1. Grotthus mechanism for proton exchange in aqueous solution.

Both the Brønsted–Lowry and Usanovich concepts attribute the special character of the proton to its small size and high mobility. However, the high charge density of the proton precludes its independent existence in solution, and solvated protons are both large and slow moving. Nevertheless, the mobility of protons appears to be very high in aqueous solution. This apparent contradiction may be explained by a three-step mechanism first proposed by Grotthus in 1806 and subsequently expanded by Huckel[38–40] (Figure 5.1), in which it is believed that the orientation step, that is, water molecule rotation, is rate determining.

A similar mechanism has been proposed for electron exchange between Fe^{2+} and Fe^{3+} in aqueous solution:[41]

Basic solution:

$$Fe^{*2+} + nH_2O + Fe(OH)^{2+}$$

$$\rightleftharpoons \left[Fe^{*} \left(\begin{array}{c} O-H \cdots\triangleright \\ | \\ H \end{array} \quad \begin{array}{c} O-H \cdots\triangleright \\ | \\ H \end{array} \right)_n \cdots\triangleright \begin{array}{c} OFe \\ | \\ H \end{array} \right]^{4+} \ddagger$$

$$\rightleftharpoons Fe^{*}(OH)^{2+} + nH_2O + Fe^{2+} \tag{43}$$

Acidic solution:

$$Fe^{*2+} + H_3O^{+} \rightleftharpoons Fe^{*}(H_3O)^{3+} \tag{44}$$

$$Fe^{*}(H_3O)^{3+} + nH_2O + Fe^{3+}$$

$$\rightleftharpoons \left[\begin{array}{c} H \\ / \\ FeO-H \\ \backslash \\ H \end{array} \cdots\triangleright \left(\begin{array}{c} O-H \cdots\triangleright \\ | \\ H \end{array} \quad \begin{array}{c} O-H \cdots\triangleright \\ | \\ H \end{array} \right)_n \cdots\triangleright \begin{array}{c} H \\ \backslash \\ OFe \\ / \\ H \end{array} \right]^{6+} \ddagger$$

$$\rightleftharpoons Fe^{*3+} + nH_2O + Fe(H_3O)^{3+} \tag{45}$$

The Grotthus-type mechanism for ferrous–ferric electron exchange is supported by the following evidence:[40]

1. Electron exchange does not occur in the absence of water.
2. The calculated inter-iron distance is large enough to accommodate one or more water molecules along with a negative bridging species.
3. The dielectric constant of a 0.143 M ethanolic solution of water is close to that of pure water, and not to that expected for the mixed solvent.

4. The exchange rate drops in mixed alcohol–water media, probably as a result of changes in the innermost solvation sheaths of the iron ions.
5. There is a large isotope effect in D_2O.
6. The activation energy of the electron exchange process is independent of the nature of many of the anions that catalyze it.
7. The electron exchange process occurs in ice as well as in liquid water, probably via a water bridge.

The slow step of the Fe(II)–Fe(III) exchange mechanism is attributed to processes occurring at the ends of the water molecule chain, where the terminal water molecules are also part of the innermost solvation spheres of the iron ions.

If proton and electron transfer occur by closely related mechanisms, at least in aqueous solution, then changes in solvation, temperature, state, and so on, should affect both processes similarly. Horne and Axelrod[40] offered the following evidence to support this contention:

1. The isotope effect of the specific conductance of water in both the liquid and solid states is approximately equal to the isotope effect of the ferrous–ferric electron exchange:

$$\frac{\kappa_{H_2O}}{\kappa_{D_2O}} \simeq \frac{k_{H_2O}}{k_{D_2O}} \tag{46}$$

   where κ is the specific conductance and k is the electron exchange rate constant.
2. The rate of proton transfer in ice is approximately the same as in water. This is also true for the electron exchange process.
3. The activation energies of proton transfer and Fe^{2+}–Fe^{3+} electron exchange in water have approximately the same temperature dependence.
4. The activation energy of the electron exchange process exceeds that of proton transfer at a given temperature in liquid water. Horne and Axelrod accounted for this discrepancy by pointing out that the solvation of the terminal water molecules of a Grotthus chain by the iron ions renders these terminal water molecules less free to rotate, which is closer to the situation existing in ice than to that in the liquid state. A comparison of the electron exchange activation energy in water and the activation energy of proton transfer in ice reveals that they are of similar magnitude.

It may therefore be concluded that proton and electron transfer occur via similar mechanisms in water. If Brønsted's view of the proton as a positive electron or an electron deficiency is adopted, the validity of Usanovich's inclusion of oxidation–reduction processes in acid–base chemistry from the point of view of the protonic theory becomes evident.

Charlot, Wolff, and LaCroix[42,43] asserted that different reaction types can be interpreted by parallel reasoning, and generalized the proton exchange basis of the Brønsted–Lowry theory to "particle exchange":

$$CD + M \rightleftharpoons C + MD \tag{47}$$

in which the species C and M compete for particle D. If D is an electron, equation (47) is an oxidation–reduction process. If the particle is a proton, equation (47) is a protolysis reaction. D may also be a polar molecule or an ion, in which case equation (47) represents a Lewis acid or base displacement, for example,

$$Hg(SCN)_3^- + Fe^{3+} \rightleftharpoons Hg(SCN)_2 + Fe(SCN)^{2+} \tag{48}$$

$$Ag(NH_3)_2^+ + 2CN^- \rightleftharpoons Ag(CN)_2^- + 2NH_3 \tag{49}$$

The results of this approach are similar to those of Usanovich in that a great deal of chemistry is systematized. According to Charlot, Wolff, and LaCroix, orders of strength depend on the particle exchanged, solvent effects may vary from inertness to simple solvation to participation in reactions, and special factors entering into reactions depend on the distinctive nature of the particle. The linkages between oxidation–reduction and acid–base phenomena, and between these reaction types and complex formation, solubility, and so on, may be accounted for by reactions incorporating more than one type of particle exchange, for example,

$$MnO_4^- + 8H^+ + 5e^- \rightleftharpoons Mn^{2+} + 4H_2O \tag{50}$$

SOLVENT SYSTEMS THEORY

The early formulations of the solvent systems theory did not consider the possibility of linkage between acid–base and oxidation–reduction reactions, despite the fact that some oxidizing and reducing agents increase solvent cation and anion concentrations, respectively, for example, equations (19) and (20). However, it was suggested that the acid–base phenomena observed in aprotic solvents such as SO_2 and $COCl_2$ are actually electron transfer (redox) processes.[17]

Ebert and Konopik[2,44] split acids and bases into donor and acceptor categories. Donor acids release solvent cations or other acids upon dissociation, and acceptor acids combine with solvent anions or other bases. Donor bases release solvent anions or other bases, and acceptor bases combine with solvent cations or other acids. The resemblance of these definitions to those of Usanovich may be noted, but again no attempt was made to link acid–base and oxidation–reduction processes.

Gutmann and Lindqvist[2,45] contended that their ionotropy concept, according to which an acid is a cation donor or anion acceptor, and a base an anion donor or cation acceptor, is virtually identical to the Usanovich definitions except in three aspects: ionotropy is linked to reactions occurring in a solvent; it excludes polyatomic ions, for example, CH_3^+, from consideration as a transferable species; and it does not generally recognize oxidation–reduction processes as acid–base reactions, although Gutmann and Lindqvist conceded that the distinction between the two becomes vague if the acidic and basic ions of a solvent are comprised of different oxidation states of the same atom; for example,

$$2I_2 \rightleftharpoons I^+ + I_3^- \tag{51}$$

is considered to represent the autoionization reaction of an iodotropic solvosystem.

LEWIS THEORY

Usanovich did not compare his concept to the electronic theory, but it seems that both are fundamentally similar in their assessments of the basis for acidity and basicity. The ability to accept an electron pair, deemed by Lewis to indicate the presence of acidic properties, is related to Usanovich's coordinate electropositive unsaturation. Similarly, electron-pair-donating tendencies can be related to coordinate electronegative unsaturation. The major difference between the classical Lewis and Usanovich acid–base theories is the inclusion of oxidation–reduction in the latter. It is therefore of interest to consider the relationship between acid–base and oxidation–reduction processes from the viewpoint of the electronic theory.

The classical Lewis formulation distinguishes between partial electron pair transfer (acid–base) and complete electron transfer (oxidation–reduction), but does not explain the increasing Lewis acidity associated with increasing oxidation state, for example,

$$\underset{\text{Weak acid}}{Fe^{2+}} \rightleftharpoons \underset{\text{Strong acid}}{Fe^{3+}} + e^- \tag{52}$$

$$\underset{\text{Weak acid}}{Sn^{2+}} \rightleftharpoons \underset{\text{Strong acid}}{Sn^{4+}} + 2e^- \tag{53}$$

$$\underset{\text{Base}}{2I^-} \rightleftharpoons \underset{\text{Acid}}{I_2} + 2e^- \tag{54}$$

the general validity of which is confirmed by Cartledge's ionic potential calculations,[46,47] as well as by experiment.[28,48] It may be noted that some of the strongest nonionic Lewis acids contain atoms of high positive oxidation state, for example, hexavalent sulfur in SO_3 and pentavalent antimony in $SbCl_5$. In addition, a sharp delineation does not always exist between Lewis acid–base and oxidation–reduction reactions, for example, as in the case of the formation of pyridine oxide[6]:

$$C_5H_5N\!:+\ddot{\underset{..}{O}}\!: \rightleftharpoons C_5H_5N\!:\ddot{\underset{..}{O}}\!: \tag{55}$$

Lewis theory, like Brønsted–Lowry theory, can account for the effect of acidity and basicity on redox potentials. The previous description of the decrease in the Fe(III)–Fe(II) reduction potential in the presence of different hydrogen acids may be interpreted in the Lewis sense as the result of more stable adduct formation between Lewis bases (perchlorate, chloride, sulfate, nitrate, and phosphate) and the Lewis acid Fe^{3+} than between the same bases and Fe^{2+}; that is, the stronger acid is the species of higher oxidation state. Gutmann[49,50] found that the oxidizing power of metal ions decreases in solvents of increasing Lewis basicity; standard reduction potentials and cathodic half-wave potentials become more negative, that is, metal ion reduction is harder to achieve, because of coordination with the solvent. Such observations agree with predictions if solvent coordination is viewed as partially providing the metal ion with the electrons it otherwise obtains only by causing the oxidation of another species. Metal ion reduction potentials rise in acidic solvents not only as a result of less metal ion–solvent coordination, but also because the solvent is able to compete with the metal ion for the solute anion, increasing the availability of the latter as an oxidizing agent and consequently increasing its reduction potential. Similarly, ligand oxidation potentials drop and anodic half-wave potentials become more positive, that is, ligand reducing power decreases in solvents of increasing Lewis acidity, since these solvents compete for ligand electrons. More basic solvents eliminate the ability of acidic species to interfere with ligand reducing power, thus increasing ligand oxidation potential. Therefore solvent basicity is associated with increasing reducing power and decreasing oxidizing power, whereas solvent acidity leads to increasing oxidizing power and decreasing reducing power, in agreement with Bancroft's[28] conclusions.

There are also instances, however, in which the reduction potential of a metal ion increases in the presence of a Lewis base. This occurs if the reduced form of the metal ion forms a more stable adduct with the base than does the oxidized form; for example, Fe^{2+} forms a more stable complex with *o*-phenanthroline than Fe^{3+}, and Cu^{+} coordinates with chloride ion to a greater extent than does Cu^{2+}.[51] The arrangement of Lewis acids (or bases) according to a single, invariant order of strength would be possible if such behavior did not exist. The occurrence of both decreased and increased redox potentials associated with complexation is another manifestation of the fact that reactivity depends not only on Lewis acid–base strength in its narrowest definition, but also on other factors, for example, polarizability, extent of ionic and covalent bonding tendencies, π-bond formation, backbonding, and so on.

The Edwards equation[52,53],

$$\log\left(\frac{K}{K_0}\right) = \alpha E_n + \beta H \tag{56}$$

where K is a rate or equilibrium constant, K_0 is this same constant in water at 25°C, H is a proton basicity factor,

$$H = \mathrm{p}K_a + 1.74 \tag{57}$$

E_n is a nucleophilicity factor,

$$E_n = E^0_{\mathrm{ox}} + 2.60 \tag{58}$$

and α and β are acid-dependent constants, recognizes the link between acid–base and oxidation–reduction reactions by attributing total reactivity to a mixture of both types of behavior. Pearson's theory of hard and soft acids and bases,[53] a qualitative descendant of the Edwards equation, states that hardness increases with increasing oxidation state and softness increases with decreasing oxidation state.

Charge-transfer spectra of transition metal complexes (which are Lewis acid–base adducts) are sometimes referred to as "redox spectra."[54] A comparison of the spectral charge-transfer band positions for complexes of a given ligand with different metal ions reveals that the energy of this band decreases with increasing metal ion oxidizing power. Conversely, for a given metal ion complexed with different ligands, the higher the reducing power of the ligand, the lower is the charge-transfer band energy.[5] It is evident that some degree of correlation of Lewis acid reactivity with oxidizing power and base reactivity with reducing power is possible, in agreement with the Usanovich view of redox processes as a part of acid–base chemistry and of redox potentials as a means for assessing relative acid–base strength.

Gutmann[50,55] distinguished between coordinating (Lewis acid–base) and redox solvents in his investigations of the effects of solvent acid–base properties on reactions in solution, but he noted similarities in the behavior of electron-pair-donating (basic) and electron-donating (reducing) solvents. The former abet the ionization of covalent solutes by coordinating with the positive part of the solute. Further solvation stabilizes the cations formed. The cations produced by the ionization of solutes in reducing solvents, for example, liquid sodium, are also stabilized by solvation:

$$H_2O + 4Na \rightleftharpoons 4Na^+ + O^{2-} + 2H^- \quad (59)$$

$$Cl_2 + 2Na \rightleftharpoons 2Na^+ + 2Cl^- \quad (60)$$

The behavior of acidic solvents, in which solute anions are stabilized by coordination and solvation, and the parallel behavior of oxidizing solvents, in which solute anions are also stabilized by solvation, for example, in liquid iodine,

$$2Na + 3I_2 \rightleftharpoons 2Na^+ + 2I_3^- \quad (61)$$

are analogous to that of basic and reducing solvents, respectively.

All of the examples cited above reveal a link between Lewis acid–base chemistry and oxidation–reduction that was acknowledged in an early presentation of the electronic theory but was not interpreted as anything more than a formal similarity.[7,58] This approach may be better understood by comparison to an earlier classification scheme that includes acids, bases, oxidizing agents, and reducing agents.

Usanovich generalized acid–base chemistry to include oxidation–reduction, but it is conceivable that the generalization could be the other way around; that is, redox chemistry may be generalized to include acid–base processes. This approach stems from the work of Ingold,[56,57] who viewed chemical reactions as electrical transactions arising from the electrophilicity, or electron-seeking tendency, of one reactant, and the nucleophilicity, that is, nucleus-seeking tendency, of the other. Ingold proposed an extended classification of chemical species based on generalized oxidation–reduction considerations.

Generalized oxidation–reduction reactions involve the donation of electrons by a nucleophile to an electrophile. Sometimes complete electron transfer occurs in such an interaction, in which case the process is an example of what is commonly regarded as a redox reaction, for example,

$$Fe + Cu^{2+} \rightleftharpoons Fe^{2+} + Cu \quad (62)$$

Not all common redox reactions are as clear-cut as equation (62) insofar as the actual transfer of electrons is concerned. For example, consider the reaction

$$I^- + Cl_2 \rightleftharpoons ICl + Cl^- \tag{63}$$

for which there may be more than one possible path for electron transfer. The iodide may reduce one atom of a homolytically cleaved chlorine molecule and then, as an iodine atom, combine with the remaining chlorine. On the other hand, the chlorine molecule may split heterolytically, leaving a Cl^+ fragment to combine with iodide. Electron transfer takes place from iodide to chloride in the former case, while in the latter the transfer occurs between chlorine atoms. A third possibility is simultaneous iodide addition and chloride expulsion, in which case two electron transfers occur.

Ingold also viewed reactions in which only a share in electrons is offered by the nucleophile to the electrophile as part of a generalized oxidation–reduction classification, justifying the inclusion of these processes, whose redox nature is somewhat vague, by pointing out what he considered to be comparable ambiguities in electron transfer in "traditional" oxidation–reduction reactions, for example, equation (63). Thus

$$H_3O^+ + OH^- \rightleftharpoons H_2O + HOH \tag{64}$$

may be considered an oxidation–reduction process if hydroxide ion is regarded as becoming formally electrically neutral, an effect comparable to the result of using hydroxide as a reducing agent:

$$OH^- \rightleftharpoons OH + e^- \tag{65}$$

Likewise, the proton transferred in equation (64) also attains formal electrical neutrality, which may be compared to its fate as an oxidizing agent:

$$H^+ + e^- \rightleftharpoons \tfrac{1}{2}H_2 \tag{66}$$

An even vaguer example of a generalized redox process, according to Ingold, is

$$F^- + BF_3 \rightleftharpoons BF_4^- \tag{67}$$

Ingold was consequently able to classify reagents as electrophiles or nucleophiles. The former are able to completely or partially accept electrons, and include both oxidizing agents and acids, for example, Cu^{2+}, Cl_2, hydrogen acids, $S_2O_8^{2-}$, some of which can behave as either or both, depending upon the circumstances, for example, equation (19). Nucleophiles are capable of complete or partial electron donation, and include both reducing agents and bases, for example, Sn^{2+}, Fe, NH_3, I^-, which, depending upon the nucleophile and upon circumstances, are able to manifest either reducing or basic behavior, or both. Ingold also recognized that some substances are capable of exhibiting both electrophilic and nucleophilic behavior, that is, amphoterism. Stannous and ferrous ions are amphoteric species in this classification.

Luder and Zuffanti[7,8,58,59] also recognized a relationship between oxidation–reduction and Lewis acid–base chemistry. They retained Ingold's classification but substituted "electrodotic" (electron donating) for "nucleophilic" (nucleus seeking), because the former term presents a more realistic description of reducing agent action, for example, sodium donates electrons but cannot be considered to be nucleus seeking. Unlike Ingold, whose classification did not distinguish between acids and oxidizing agents, or between bases and reducing agents, Luder and Zuffanti maintained that a distinction does exist, despite cases of apparent overlap, for example, see equations (19), (20), (32), and (33). They contended that instances of simultaneous acid and oxidizing action, or simultaneous basic and reducing action, are limited to processes involving the formation or reaction of the solvent ions of amphoteric solvents, and are consequently not generally characteristic of acids and bases. Many strong acids, for example, SO_3, have no oxidizing power, and many strong oxidizing agents, for example, MnO_4^-, evidence no acidity (although there is an argument, to be discussed later, supporting the acidity of permanganate).

The classification of a reagent as electrophilic or electrodotic depends upon the reaction under consideration, that is, amphoterism is common in both acid–base and redox processes. For example, HCl is electrodotic when it acts as an acid towards water or as an oxidizing agent towards metals; it is electrophilic when it behaves as a base towards SO_3 or as a reducing agent towards permanganate.

Lewis theory, according to Luder and Zuffanti, thus distinguishes between three reaction types:

1. *Acid–base reactions* Those involving electron pair sharing.
2. *Redox reactions* Those involving complete electron transfer.
3. *Free radical reactions* Odd electron processes.

The acid–base processes are often referred to as "closed shell–closed shell" interactions to distinguish them from the other types ("open shell–open shell" interactions).[60]

According to the principle that the acidity of an element increases with increasing oxidation state, it might be expected that permanganate, which contains heptavalent manganese, should be one of the strongest acids. As mentioned previously, however, Luder and Zuffanti cited MnO_4^- as a prime example of a strong oxidizing agent that manifests no acidity, in support of their contention that acid–base and oxidation–reduction processes are inherently different. If MnO_4^- is regarded to be composed of Mn^{7+} and four oxide ions, part of the apparent lack of acidity may be due to the presence of the latter, which are basic. It must be noted at this point, however, that SO_3 and CO_2 are not subject to the loss of their predominant acidity simply because oxygen is present, despite the fact that their electropositive atoms are of lower valence than manganese in MnO_4^-. The possibility of coordinate saturation of manganese, similar to that of carbon in the orthocarbonic acid anion, CO_4^{4-}, or that of sulfur in the sulfate ion, may be considered as an explanation of the lack of apparent acidity. Permanganate is converted to MnO_3^+ ion in aqueous H_2SO_4; the visible and ultraviolet spectra of this species are consistent with a trigonal planar structure.[61] By analogy to other trigonal planar, coordinately unsaturated substances, for example, SO_3, CO_3^{2-}, and BCl_3, which become tetrahedral and coordinately saturated upon the formation of an adduct with a base, permanganate may be regarded as a tetrahedral species no longer susceptible to further coordination, that is, a species that has no apparent acidity. The electron dot structure of the ion would seem to be in accord with such an interpretation.

Usanovich[62] challenged the assertion of Luder and Zuffanti, insisting that oxidizing agents are equivalent to acids. Permanganate, according to this viewpoint, is an acid as well as an oxidizing agent because it causes the disappearance of bases (many of which are also good reducing agents). As supporting evidence Usanovich cited the work of Duke,[63] who believed that ions of like charge cannot react directly with each other, but must interact through neutral or oppositely charged intermediates. Examples of such behavior are found in both acid–base and oxidation–reduction processes; for example, Duke maintained that the conversion of dihydroxotetraaquoiron (III) to hydroxopentaaquoiron (III) does not occur via the prior dissociation of the solvent in aqueous solution,

$$[Fe(OH)_2(H_2O)_4]^+ + H^+ \rightleftharpoons [Fe(OH)(H_2O)_5]^{2+} \tag{68}$$

because the close approach of two positively charged species is required. Instead, the dihydroxo complex abstracts a proton directly from electrically neutral water:

$$[Fe(OH)_2(H_2O)_4]^+ + H_2O \rightleftharpoons [Fe(OH)(H_2O)_5]^{2+} + OH^- \tag{69}$$

The oxidation of stannous ion by ferric ion does not occur in the absence of a relatively large HCl concentration [equations (23) and (24)] because both reactants are positively charged and require a large chloride ion concentration, which permits the formation of metal–chloride complexes. The rate of this reaction decreases when $HClO_4$ is substituted for HCl, because perchlorate is not as good a ligand as chloride ion.

Mechanisms of processes involving permanganate oxidation are particularly difficult to explain, according to Duke, because there is often no obvious positive or neutral intermediate, as, for example, in the alkaline oxidation of sulfites, phosphites, or formates. The constituents common to such processes are, in addition to the reducing agent and permanganate, water and alkali or alkaline-earth metal cations. The cations are positively charged but are unlikely intermediates, because such an assumption implies their oxidation to higher oxidation states by permanganate, which is only a remote possibility in view of their closed-shell electronic configurations. Duke therefore suggested that the first step in MnO_4^- oxidation, at least in water, involves an equilibrium between permanganate, water, the manganate ion MnO_4^{2-} (a species stable only in strongly alkaline solutions or in basic melts) and the hydroxyl radical, which serves as the neutral intermediate that actually oxidizes the reducing agent:[62–64]

$$MnO_4^- + H_2O \rightleftharpoons MnO_4^{2-} + OH + H^+ \tag{70}$$

Usanovich cited the formation of protons in equation (70) as evidence supporting the simultaneous acidity and oxidizing power of permanganate.

The validity of equation (70) is questionable because it requires a reverse process that constitutes a trimolecular reaction between three species of low concentration, and an alternate formulation is considered more likely:[64–67]

$$MnO_4^- + OH^- \rightleftharpoons MnO_4^{2-} + OH \tag{71}$$

We note at this point that a Grotthus-type electron transfer mechanism analogous to that postulated for aqueous ferrous–ferric ion electron exchange, as well as for aqueous proton transfer, would seem to be a logical possibility for equations (70) and (71) [especially if coupled with proton transfer in equation (70)], emphasizing again the link between acid–base and oxidation–reduction chemistry in general, and between acids and permanganate in particular.

The alkaline oxidation of cyanide to cyanate by MnO_4^- is believed to occur via a mechanism incorporating equation (71) or a similar equilibrium between cyanide and MnO_4^-.[68] Ladbury and Cullis[64] stated that the two predominant mechanisms of permanganate oxidation, at least in alkaline aqueous solution, are electron abstraction [equation (70) or (71)] followed by hydroxyl radical oxidation of the reducing agent, and, in the case of protonic reducing agents, proton abstraction by hydroxide,

$$HA + OH^- \rightleftharpoons A^- + H_2O \tag{72}$$

followed by reaction of the base A^- with MnO_4^-, also possibly via a Grotthus mechanism. Permanganate causes the disappearance of a base in either case.

Equilibria of the form of equation (71), be it with the base OH^-, CN^-, or any other base A^-, do not constitute the only mechanistic evidence in support of permanganate acidity. As far back as 1922, Holluta[69–71] suggested the following steps for the oxidation of formic acid by MnO_4^- in aqueous solutions of varying acidity:

Weak acid solution:

$$MnO_4^- + HCOO^- \rightarrow MnO_4^{3-} + CO_2 + H^+ \tag{73}$$

Neutral solution:

$$MnO_4^- + HCOO^- + H_2O \rightarrow MnO_4^{3-} + CO_3^{2-} + 3H^+ \tag{74}$$

Basic solution:

$$MnO_4^- + HCOO^- + OH^- \rightarrow MnO_4^{2-} + CO_3^{2-} + H_2O \tag{75}$$

It must be noted at this point that equations (70)–(75) all involve the formation or reaction of solvent ions of an amphoteric solvent, which, as previously mentioned, is insufficient as a general criterion for the identification of acids and bases in the opinion of Luder and Zuffanti.[7,58] However, Wiberg and Stewart[72] argued that equations (73) and (74) involve the transfer of either a hydride ion or an electron pair from formate to permanganate:

$$({}^-O)(O{=})C{-}H + O{-}MnO_3^- \rightarrow CO_2 + HMnO_4^{2-} \tag{76}$$

$$H{-}C({=}O)(O^-) + O{-}MnO_3^- \rightarrow H^+ + CO_2 + MnO_4^{3-} \tag{77}$$

Equation (76) can be regarded as a Lewis acid displacement in which permanganate displaces carbon dioxide from the hydride ion. Since Wiberg and Stewart specify the transfer of an electron pair (as opposed to two individual electrons) in equation (77), it is possible to view this process as the formation and subsequent dissociation of an unstable Lewis acid–base adduct.

The above-cited mechanistic evidence indicates that a case can be made in support of Usanovich's assertion that permanganate specifically is both an oxidizing agent and an acid, despite its apparent lack of acidity. By extension, all oxidizing agents are equated with acids, and reducing agents with bases.

The quantum mechanical formulations of the Lewis theory are more inclusive than the Luder–Zuffanti viewpoint. The Mulliken treatment,[60,73–75] based on the contributions of a "no-bond" state $\Psi(A,B)$, comprised mainly of electrostatic interactions, and a state of complete charge transfer (covalent interaction) $\Psi(A^-B^+)$ to the actual state Ψ_{AB} of a Lewis acid–base adduct,

$$\Psi_{AB} = a\Psi(A,B) + b\Psi(A^-B^+) \tag{78}$$

(where a and b are weighting coefficients), does not specify that the transfer involves pairs of electrons. Equation (78) also permits all degrees of donation, including the complete electron transfer characteristic of oxidation–reduction ($b^2/a^2 = \infty$). Although Mulliken originally developed his treatment for the class of weak Lewis acid–base adducts known as charge-transfer complexes, he subsequently realized its general applicability, and explicitly included oxidation–reduction processes in his formulation, as well as closed shell–closed shell interactions. The Klopman–Hudson reactivity treatment,[60,76–78] also based on weak charge transfer, can, in principle, also be extended to redox reactions.

Jensen[60] maintained that oxidizing power (complete acceptance of electrons), acidity (acceptance of a share in an electron pair), basicity (donation of a share in an electron pair), and reducing power (complete electron donation) may be regarded as degrees along a continuum instead of as separate phenomena. Along this continuum there are gray areas in which these processes overlap; for example, the reaction

$$SO_3^{2-} + ClO_3^- \rightleftharpoons SO_4^{2-} + ClO_2^- \tag{79}$$

may be regarded as either a redox reaction or as a Lewis base displacement of ClO_2^- from an acidic oxygen atom by SO_3^{2-}.

Although it has been argued that the quantum mechanical treatments blur the distinction between two different types of phenomena (acid–base and oxidation–reduction),[14] one of the results of this broader view of the electronic theory is the qualitative equating of the Lewis and Usanovich acid–base concepts. However, oxidizing and reducing agents are explicitly designated as acids and bases, respectively, only in the latter approach.

5.3 CONCLUSIONS

The evolution of the acid–base concept since Boyle's initial seventeenth century attempt to explain the properties of acids was, at first, a shrinking process, continually restricting the number of substances regarded as acids and bases, and culminated in the narrow-based Arrhenius theory in the late nineteenth century. The twentieth century has witnessed a reversal in this trend. Successive acid–base concepts have tended toward the opposite direction for over half a century; that is, the range of processes regarded as acid–base reactions has grown to include a wide variety of chemical reactions, and has contributed greatly to the knowledge of the fundamental properties of matter.

REFERENCES

1. Usanovich, M. I., *J. Gen. Chem. USSR*, **9**, 182 (1939).
2. Gyenes, I., "Titration in Nonaqueous Media," Van Nostrand, Princeton, NJ, 1967.
3. Brønsted, J. N., *Ber.*, **61**, 2049 (1928).
4. Hall, N. F., *J. Chem. Educ.*, **17**, 124 (1940).
5. Day, Jr., M. C., and Selbin, J., "Theoretical Inorganic Chemistry," 2nd ed., Reinhold, New York, 1969.
6. Huheey, J. E., "Inorganic Chemistry, Principles of Structure and Reactivity," Harper and Row, New York, 1972.
7. Luder, W. F., and Zuffanti, S., "The Electronic Theory of Acids and Bases," 2nd rev. ed., Dover, New York, 1961.
8. Luder, W. F., *Chem. Rev.*, **27**, 547 (1940).
9. Brønsted, J. N., *Rec. Trav. Chim. Pays-Bas*, **42**, 718 (1923).
10. Brønsted, J. N., *Chem. Rev.*, **5**, 231 (1928).
11. Alyea, H. N., *J. Chem. Educ.*, **18**, 206 (1941).
12. Bates, R. G., "Treatise on Analytical Chemistry," Part I, Vol. I, Kolthoff, I. M., Elving, P. J., and Sandell, E. B., Eds., Wiley-Interscience, New York, 1959, Chap. 10.
13. Bates, R. G., "Determination of pH, Theory and Practice," 2nd ed., Wiley, New York, 1973.
14. Jørgensen, C. K., *Chimia*, **28**, 605 (1974).

15. Schwarzenbach, G., *Helv. Chim. Acta*, **13**, 870 (1930).
16. Kolthoff, I. M., "Treatise on Analytical Chemistry," Part I, Vol. I, Kolthoff, I. M., Elving, P. J., and Sandell, E. B., Eds., Wiley-Interscience, New York, 1959, Chap. 11.
17. Bradley, R. S., *J. Chem. Educ.*, **27**, 208 (1950).
18. Lewis, G. N., *J. Franklin Inst.*, **226**, 293 (1938).
19. Weiss, J., *Naturwissenschaften*, **23**, 64 (1935).
20. Weiss, J., *J. Chem. Soc.*, 309 (1944).
21. Duke, F. R., "Treatise on Analytical Chemistry," Part I, Vol. I, Kolthoff, I. M., Elving, P. J., and Sandell, E. B., Eds., Wiley-Interscience, New York, 1959, Chap. 15.
22. Ball, T. R., Wulfkuehler, W., and Wingrad, S., *J. Am. Chem. Soc.*, **57**, 1729 (1935).
23. Brønsted, J. N., *Z. Phys. Chem.*, **169**, 52 (1934).
24. Brønsted, J. N., *J. Phys. Chem.*, **30**, 777 (1926).
25. Mohanty, J. G., and Chakravorty, A., *Inorg. Chem.*, **16**, 1561 (1977).
26. Audrieth, L. F., and Schmidt, M. T., *Proc. Nat. Acad. Sci. U.S.A.*, **20**, 221 (1934).
27. Audrieth, L. F., and Moeller, T., *J. Chem. Educ.*, **20**, 219 (1943).
28. Bancroft, W. D., *J. Chem. Educ.*, **11**, 267 (1934).
29. Bates, R. G., "Treatise on Analytical Chemistry," Part I, Vol. I, Kolthoff, I. M., Elving, P. J., and Sandell, E. B., Eds., Wiley-Interscience, New York, 1959, Chap. 9.
30. Kraus, C. A., *J. Am. Chem. Soc.*, **29**, 1557 (1907).
31. Kraus, C. A., *ibid.*, **30**, 1323 (1908).
32. Kraus, C. A., *ibid.*, **36**, 864 (1914).
33. Kraus, C. A., *ibid.*, **43**, 749 (1921).
34. Hart, E. J., *Acc. Chem. Res.*, **2**, 161 (1969).
35. Rabani, J., "Solvated Electron," Advances in Chemistry Series 50, American Chemical Society, Washington, D.C., 1965, Chap. 17.
36. Anbar, M., *ibid.*, Chap. 6.
37. Matheson, M. S., *ibid.*, Chap. 5.
38. Grotthus, C. J. T., *Ann. Chem.*, **58**, 54 (1806).
39. Huckel, E., *Z. Elektrochem.*, **34**, 546 (1928).
40. Horne, R. A., and Axelrod, E. H., *J. Chem. Phys.*, **40**, 1518 (1964).
41. Reynolds, W. L., and Lumry, R. W., *ibid.*, **23**, 2460 (1955).
42. Charlot, G., Wolff, J. P., and LaCroix, S., *Anal. Chim. Acta*, **1**, 73 (1947).
43. Charlot, G., Wolff, J. P., and LaCroix, S., *ibid.*, **3**, 285 (1949).
44. Ebert, L., and Konopik, N., *Oesterr. Chem. Ztg.*, **50**, 184 (1949).
45. Gutmann, V., and Lindqvist, I., *Z. Phys. Chem.*, **203**, 250 (1954).
46. Cartledge, G. H., *J. Am. Chem. Soc.*, **50**, 2855 (1928).
47. Cartledge, G. H., *ibid.*, **50**, 2863 (1928).
48. Campbell, J. A., *J. Chem. Educ.*, **23**, 525 (1946).
49. Gutmann, V., *Fortschr. Chem. Forsch.*, **27**, 59 (1972).
50. Gutmann, V., *Pure Appl Chem.*, **27**, 73 (1971).
51. Ringbom, A., "Treatise on Analytical Chemistry," Part I, Vol. I, Kolthoff, I. M., Elving, P. J., and Sandell, E. B., Eds., Wiley-Interscience, New York, 1959, Chap. 14.
52. Edwards, J. O., *J. Am. Chem. Soc.*, **76**, 1540 (1954).

53. Pearson, R. G., *ibid.*, **85**, 3533 (1963).
54. Kettle, S. F. A., "Coordination Compounds," Appleton-Century-Crofts, New York, 1969.
55. Gutmann, V., *Chem. Br.*, **7**, 102 (1971).
56. Ingold, C. K., *J. Chem. Soc.*, 1120 (1933).
57. Ingold, C. K., *Chem. Rev.*, **15**, 225 (1934).
58. Luder, W. F., *J. Chem. Educ.*, **19**, 24 (1942).
59. Luder, W. F., *ibid.*, **22**, 301 (1945).
60. Jensen, W. B., *Chem. Rev.*, **78**, 1 (1978).
61. Cotton, F. A., and Wilkinson, G., "Advanced Inorganic Chemistry," 2nd ed., Wiley-Interscience, New York, 1966.
62. Usanovich, M. I., *J. Gen. Chem. USSR*, **21**, 2181 (1951).
63. Duke, F. R., *J. Am. Chem. Soc.*, **70**, 3975 (1948).
64. Ladbury, J. W., and Cullis, C. F., *Chem. Rev.*, **58**, 403 (1958).
65. Stamm, J., *Z. Angew. Chem.*, **47**, 791 (1934).
66. Symons, M. C. R., *J. Chem. Soc.*, 3956 (1953).
67. Symons, M. C. R., *ibid.*, 3676 (1954).
68. Freund, T., *J. Inorg. Nucl. Chem.*, **15**, 371 (1960).
69. Holluta, J., *Z. Phys. Chem.*, **101**, 34 (1922).
70. Holluta, J., and Weiser, N., *ibid.*, **101**, 489 (1922).
71. Holluta, J., *ibid.*, **102**, 276 (1922).
72. Wiberg, K., and Stewart, R., *J. Am. Chem. Soc.*, **78**, 1214 (1956).
73. Mulliken, R. S., *J. Chem. Phys.*, **19**, 514 (1951).
74. Mulliken, R. S., *J. Phys. Chem.*, **56**, 801 (1952).
75. Mulliken, R. S., *J. Am. Chem. Soc.*, **74**, 811 (1952).
76. Hudson, R. F., and Klopman, G., *Tetrahedron Lett.*, **12**, 1103 (1967).
77. Klopman, G., and Hudson, R. F., *Theor. Chim. Acta*, **8**, 165 (1967).
78. Klopman, G., *J. Am. Chem. Soc.*, **90**, 223 (1968).

CHAPTER

6

PROCEDURES FOR THE EXPERIMENTAL VERIFICATION AND RECONCILEMENT OF THE VARIOUS ACID–BASE THEORIES

Careful examination of the Brønsted–Lowry and Lewis acid–base theories reveals support for a strong connection between acid–base and oxidation–reduction processes. The Usanovich theory goes further and explicitly includes oxidation–reduction reactions in acid–base chemistry. However, it offers no experimental evidence to justify the classification of oxidizing agents as acids and reducing agents as bases, nor does it suggest a general mechanism or specific, structure-dependent phenomenon by which oxidizing and reducing agents manifest their respective acidic and basic tendencies that is comparable to protolysis in the protonic concept or to coördination in the electronic concept.

The hypothesis that oxidizing agents are acids may be tested by applying conventional methods of measuring acidity, that is, pH measurements and acid–base titrations, to the determination of the acidity of oxidizing agents. pH measurements were made on a series of aqueous solutions of constant hydrogen acid concentration and varying oxidizing agent concentration to determine the effect of the latter on solution acidity. If oxidizing agents indeed manifest acidic tendencies, then the pH should decrease as oxidant concentration increases. However, the magnitude of the pH change is expected to be small, and consequently the problem of other factors that change pH and thus interfere with the observation of oxidant acidity arises.

The increase in ionic strength with increasing oxidant concentration will result in decreased activity of all ionic species present and increased pH, in accord with the various forms of the Debye–Huckel equation. Although it was uncertain whether the Debye–Huckel effect was of sufficient magnitude to obscure pH changes resulting from the acidity of the oxidizing agent, it was certainly a complication, and therefore a series of experiments was performed to determine the pH change in a series of acid–oxidant solutions at constant ionic strength.

The change in liquid-junction potential with change in solution composition also represents a complicating factor; unlike the Debye–Huckel effect, changes in liquid-junction potential are often of unpredictable magnitude and direction because the ionic mobilities from which they are calculated are not known with certainty under all experimental conditions, for example, changing pH and changing ionic strength, for all species. The contribution of an ionic species to the liquid-junction potential depends on its activity, the sign and magnitude of its charge, and its transport number t. The transport number depends on the ionic mobilities u and activities a of all species present in solution:[9]

$$t_i = \frac{a_i u_i}{\sum_j a_j u_j} \tag{1}$$

The sum of the transport numbers of all the species in any solution must therefore be unity. The effect of changing liquid-junction potential was also eliminated in the above-mentioned pH experiment involving the measurement of pH of a series of acid–oxidant solutions at constant ionic strength, because the sum of the transport numbers of those species (hydrogen ion and the ions of the supporting electrolyte) whose activities remained constant was always $\geqslant 0.9975$ and assured negligible contributions to the liquid-junction potential from the ions of the oxidizing agent, whose activities changed with increasing oxidant concentration. In addition, the change in emf with increasing oxidant concentration was measured for a series of acid–oxidant solutions without supporting electrolyte in a cell without a salt bridge, to insure that the change in liquid-junction potential did not mask the acidity change.

Potentiometric titrations of bases with oxidizing agents were performed, since an acid may be defined, on a practical basis, as any substance capable of neutralizing a base. The results of these titrations were analyzed with regard to reproducibility, sharpness of equivalence point, and stoichiometry of the titration reaction.

The fact that the structures of many oxidants, for example, dichromate and permanganate ions, depend upon the acidity of the medium in which they are dissolved implies a connection between oxidant acidity and structure that, in turn, suggests a mechanism by which oxidizing agents manifest acidity. Paper-strip electrophoresis experiments were performed with various oxidizing agents in supporting media of varying acidity in an attempt to elucidate this mechanism and to determine the presence of any previously unobserved species whose existence could be correlated with oxidizing agent acidity.

6.1 REAGENT-PREPARATION

WATER

The high degree of accuracy required in the pH and other emf measurements necessitated high-purity water for preparation of stock and sample solutions that was free of ionic impurities that could alter ionic strength and organic contaminants capable of undergoing oxidation in the presence of strong oxidizing agents. Water of the requisite purity was prepared by passing tap water through two deionizing columns, distilling the doubly deionized effluent in a glass still over alkaline permanganate to oxidize organic substances and trap the CO_2 thus produced,[1] and, finally, redistilling. Conductance measurements indicated a conductance of approximately 1.7×10^{-6} mho for water treated in this manner, as compared to approximately 5.9×10^{-6} mho for ordinary deionized water.

ACID AND BASE STOCK SOLUTIONS

Stock solutions of HCl and $HClO_4$ were prepared by diluting the respective concentrated reagent-grade acids with water and were standardized against analytical reagent-grade sodium carbonate, utilizing both visual indicator (methyl orange) and potentiometric end-point detection techniques.[2,3]

Reagent-grade sodium hydroxide was dissolved in a small quantity of water to prepare a concentrated NaOH solution. This intermediate step in the production of a stock solution possesses an advantage over direct preparation of a stock solution of the desired concentration because the sodium carbonate impurity present in solid NaOH is only slightly soluble in a concentrated NaOH solution, but very soluble in dilute NaOH;[2,4] addition of a small quantity of barium hydroxide insures precipitation of carbonate.[5] Insoluble material was then removed by filtration, and the filtrate was transferred to a plastic bottle, in which it was immediately diluted with freshly boiled water and thus stored under a parafilm seal. The bottle was opened only to facilitate withdrawal of aliquots; on such occasions nitrogen gas was bubbled through the remainder of the contents to prevent air from entering the bottle during aliquot removal, and the bottle was then closed and resealed. Potassium hydroxide stock solutions were prepared and stored in an identical manner, except that the use of barium hydroxide to precipitate the carbonate impurity is necessary, rather than optional, in this case, since potassium carbonate is freely soluble even in concentrated KOH.[2,5] Sodium and potassium hydroxide solutions were standardized potentiometrically against primary standard potassium hydrogen phthalate.[2,4]

OXIDANT STOCK SOLUTIONS

Weighed quantities of primary standard potassium dichromate were dissolved in water to prepare dichromate stock solutions, whose concentrations were verified by gravimetric determination as barium chromate.[6,7]

Ferric chloride and ferric perchlorate stock solutions were prepared by dissolving the respective reagent-grade chemicals in water. The iron content of these solutions was determined gravimetrically by precipitation from alkaline solution as the hydrous oxide and subsequent ignition to Fe_2O_3.[2,8]

Potassium permanganate stock solutions were prepared by dissolving $KMnO_4$ in water, boiling the solution for 1 hour to oxidize organic impurities and expel CO_2, and filtering through sintered glass to remove MnO_2.[2] These solutions were stored in brown bottles in the dark under a parafilm seal, and were opened only when aliquots were removed, after which the remaining stock solution was deaerated with nitrogen gas, and the bottle closed and resealed. Permanganate solutions were standardized by titration against potassium oxalate.[2]

STOCK SOLUTIONS OF SUPPORTING ELECTROLYTE

Reagent-grade lithium chloride was dissolved in water and standardized against silver nitrate via the Mohr method of determining soluble chloride.[2–4]

Supporting-electrolyte stock solutions for use in electrophoresis experiments were prepared from aliquots of the $HClO_4$ or KOH stock solutions over a pH range from 1.1 to 12.7. Lithium or potassium perchlorate was added where required to bring the ionic strength of each supporting medium to 0.10.

SAMPLE SOLUTIONS

Acid–oxidant sample solutions for use in the pH and liquid-junction measurements were prepared by diluting mixtures of aliquots drawn from the acid and oxidant (and, where necessary, supporting electrolyte) stock solutions with water in 1-liter volumetric flasks, thus assuring the availability of replicate samples from each sample solution. The pipettes utilized to withdraw aliquots from the stock solutions were calibrated[2] so that their delivery volumes were known to $\leqslant \pm 0.01$ ml; no pipette smaller than 10 ml was used, so that the analytical concentrations of all constituents of the sample solutions were known to at least ± 1 part per thousand. The pH titrations involved direct use of stock solutions.

The following sample solutions were prepared for use in electrophoresis experiments by dissolving the appropriate amounts of the requisite reagent-grade chemicals in water: 0.05 *M* $K_2Cr_2O_7$ in 0.10 *M* $LiClO_4$ and 0.10 *M* $Fe(ClO_4)_3$ in 0.10 *M* $LiClO_4$. A solution of 0.10 *M* $KMnO_4$ in 0.10 *M* $LiClO_4$ was prepared by diluting an aliquot of stock $KMnO_4$ and adding the requisite amount of lithium perchlorate.

BUFFERS

The pH values of commercially produced buffers are not known to the degree of accuracy required for the pH measurements, that is, $\pm$ a few thousandths of a pH unit. It was therefore necessary to prepare buffer solutions of sufficient accuracy to standardize the pH meter–electrode assembly over the ranges of pH studied. Table 6.1 contains a list of the buffers utilized. Freshly boiled water, or boiled water stored under NaOH, was used in buffer preparation.

6.2 PROCEDURES AND EQUIPMENT

pH MEASUREMENTS

The pH of aliquots drawn from acid–oxidant sample solutions were measured to one thousandth of a pH unit with a Corning model 12 pH meter equipped with Corning models 476022 glass and 476029 calomel electrodes. Samples were contained in a jacketed, four-necked flask with a single gas outlet (Figure 6.1). Sample temperature was maintained at 25°C by either immersing the sample container in a constant-temperature bath or by circulating water from the bath through the flask jacket. Samples were

TABLE 6.1 Buffer Solutions Utilized for Standardization of pH Meter–Electrode Assembly in pH Experiments

Buffer	Preparation	pH at 25°C	References
0.05 *M* potassium tetroxalate (secondary standard)	Dissolution of weighed quantity of reagent-grade potassium tetroxalate.	1.679	9–12
0.01549 *M* KH_2PO_4 + 0.01221 *M* HCl	Dissolution of weighed quantity of reagent-grade potassium dihydrogen phosphate; dilution of stock HCl.	2.270	5
0.01105 *M* KH_2PO_4 + 0.008710 *M* HCl	Dissolution of weighed quantity of reagent-grade potassium dihydrogen phosphate; dilution of stock HCl.	2.366	5
0.05 *M* potassium hydrogen phthalate (primary standard)	Dissolution of weighed quantity of primary-standard-grade potassium hydrogen phthalate.	4.004	5,9–12

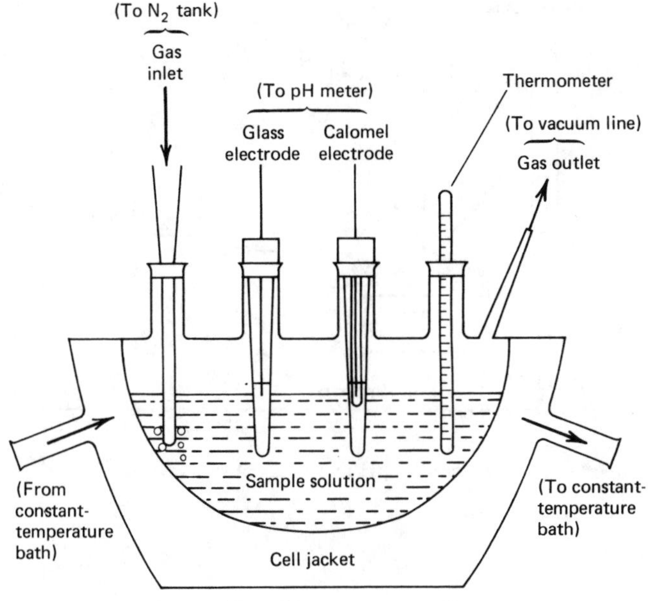

Figure 6.1. pH measurement cell.

deaerated with nitrogen gas for several minutes prior to the beginning of measurement, during which pressure buildup in the flask was prevented by mild suction.

The pH meter–electrode assembly was connected to a transformer-regulated, 120-V constant AC voltage to minimize the effect of line fluctuations. The meter was standardized daily with appropriate standard buffer solutions (Table 6.1), that is, buffers of pH approximating the sample pH, under experimental conditions identical to those of sample measurements.

The time required for the system to attain equilibrium in each measurement, as indicated by no change in pH over a 10–15 min period, was on the order of 1–2 hour. The flask and electrodes were carefully rinsed after each measurement, first with several portions of deionized, distilled water, and then with several portions of the next sample solution. Replicate measurements were obtained for each sample.

Figure 6.2 is a schematic diagram of the complete pH measurement system described above.

emf MEASUREMENTS IN A CELL WITHOUT A SALT BRIDGE

The apparatus consisted of an H-shaped cell containing a sintered glass disk that separated the solutions in the two half-cells (Figure 6.3). The reference

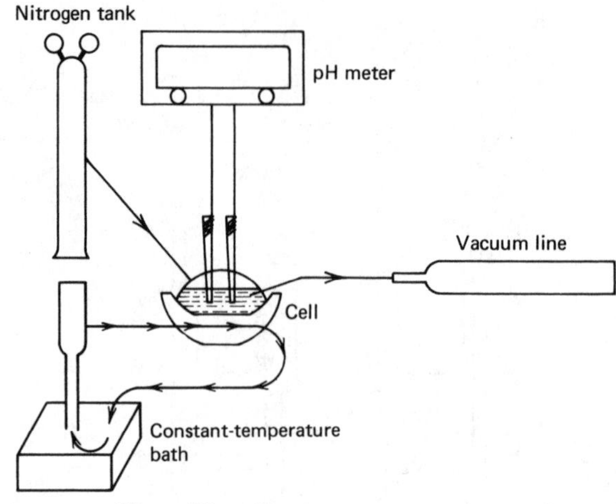

Figure 6.2. pH measurement system.

half-cell was composed of a previously unused, Fisher model 13-639-3 glass electrode with BNC connector, immersed in an aqueous solution of approximately 0.005 *M* HCl. The indicator half-cell was composed of a second new electrode of the identical type immersed in an aqueous solution of equal HCl concentration, which also contained potassium dichromate in a concentration varying from zero to approximately the same molarity as HCl. Each half-cell also contained a grounded platinum wire to minimize the effect of stray currents, since the cell potentials measured were of the order of only a few millivolts.

The high impedance of a cell containing two glass electrodes causes the output signal of the cell to be very small. It is convenient to amplify this signal and therefore an amplifier with a gain of 3, powered by a 30-V power supply, was included in the measuring circuitry. The reference half-cell was attached to the positive terminal of the amplifier, and the indicator half-cell to the negative terminal. The amplifier output was read on a Hewlett-Packard model 3430 A digital voltmeter.

The solutions in both half-cells were deaerated with nitrogen gas for several minutes prior to the beginning of each measurement, and the temperature was maintained at 25°C by immersing the cell in a constant-temperature bath.

Equilibrium, as indicated by a fluctuation in the voltmeter reading of only a few tenths of a millivolt (and therefore a fluctuation in actual junction potential one-third as large) over a 10–15 min period, was attained in each measurement after about an hour. The indicator half-cell and

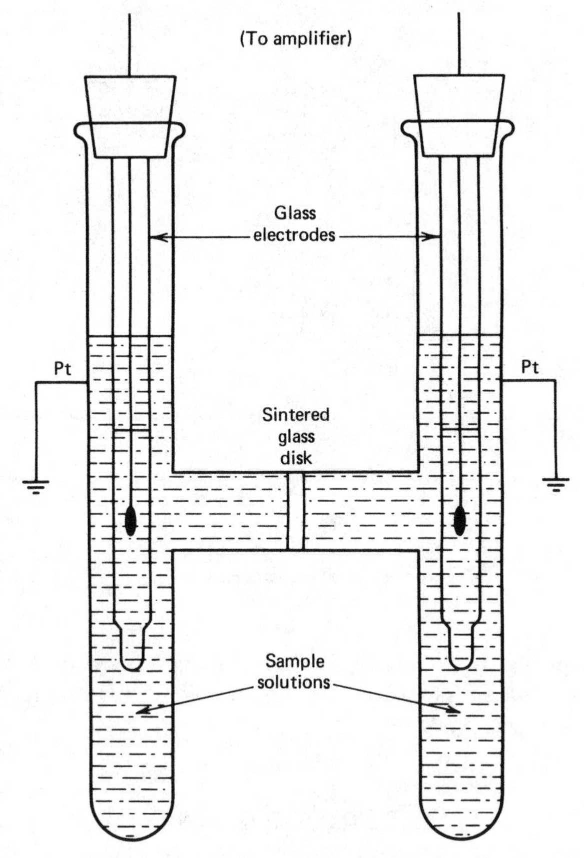

Figure 6.3. H-Shaped cell without salt bridge.

electrode were carefully rinsed after each measurement with several portions of the next sample solution. The solution in the reference half-cell was replaced by a fresh aliquot every two to three measurements. Replicate measurements were obtained for each sample solution.

Figure 6.4 is a schematic diagram of the emf measurement system described above.

pH TITRATIONS

Portions of base (either aliquots of stock NaOH or weighed Na_2CO_3 samples) were titrated potentiometrically with stock oxidizing agent solutions under the constant, slow bubbling of nitrogen gas to prevent absorption of atmospheric CO_2, and constant stirring to insure solution homogeneity. pH measurements were made to an accuracy of a hundredth

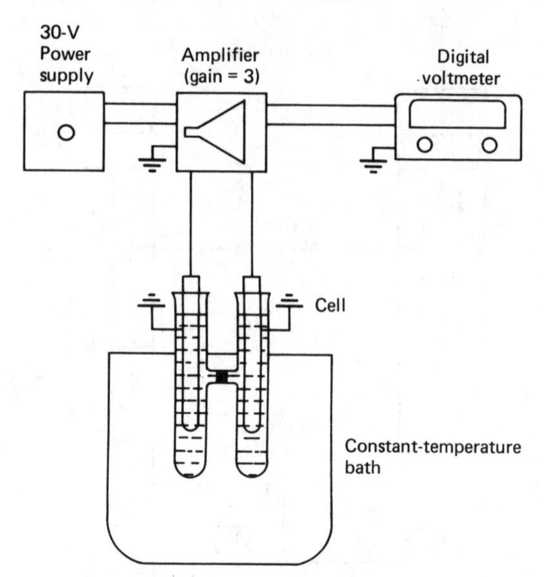

Figure 6.4. Emf measurement system.

of a pH unit during titration with the same pH meter–electrode assembly as in the static pH measurements, with a higher concentration of readings in the region of the end point.

ELECTROPHORESIS

The electrophoresis apparatus consisted of a Shandon electrophoresis bath and an adjustable DC power supply. The bath was filled with one of the previously prepared supporting electrolyte solutions. Since these supporting media were not discarded following use, but stored and reused in the study of several oxidizing agents, it was necessary to periodically check the pH of each supporting electrolyte with a Corning model 12 pH meter and Corning glass and calomel electrodes to assure constancy of pH. It was found that solutions at both ends of the pH scale ($pH<4$ and $pH>10$) maintained a fairly constant pH with time and use, that is, fluctuations were less than one-tenth of a pH unit, while those at intermediate pH did not remain constant. However, it will be shown that the wide variations in the intermediate pH range were not significant for the oxidizing agents studied.

Whatman No. 3 mm chromatographic filter paper was utilized as a chromatographic medium in most of these experiments. Initial experiments with this paper produced relatively high migration currents and a significant amount of evaporation from the supporting medium, as evidenced by excessive condensation in the upper part of the electrophoresis chamber. This problem was solved by cutting the paper into strips one-fourth their original width; decreasing the cross-sectional area in this manner decreased the current without affecting the quality of the results obtained. Mallinckrodt Chromar Sheet 500, a blend of chromatographic-grade silicic acid (H_2SiO_5, 70%) and glass fiber (30%) was used in some of the permanganate experiments instead of the Whatman paper in an attempt to avoid MnO_4^- decomposition due to the presence of organic compounds in the paper.

Each paper strip was supported by a plastic ruler to prevent sagging and to facilitate direct reading of migration distance. The ruler was suspended between the supports of the electrophoresis chamber so that the ends of the paper could dip into the supporting electrolyte. Each paper strip was equilibrated in the supporting medium for at least a half hour before application of a sample spot and initiation of an experiment to minimize migration due to solvent flow. Supporting media were periodically deaerated with nitrogen.

A circular spot was applied to the stationary phase with a 10- or 50 μl disposable pipette and its location was noted from the plastic ruler. Experiments were performed at an applied DC voltage of 100 V for one hour. Spot development was achieved by spraying the strips with an 0.2% solution of the redox indicator diphenylamine in aqueous sulfuric acid; the oxidized form of this indicator is either violet or green, depending on the exact indicator concentration.[2]

6.3 RESULTS AND DISCUSSION

DICHROMATE

The observed changes of pH with increasing oxidizing agent concentration were compared to predicted pH trends for each set of acid–oxidant solutions studied. Predicted pH values were calculated for each oxidant concentration in a series of sample solutions at fixed acid concentration in the usual manner:

$$\mathrm{pH} = -\log a_H = -\log(\gamma_H C_H) \tag{2}$$

where a_H is the hydrogen ion activity, C_H is the hydrogen ion concentration, and γ_H is the activity coefficient of hydrogen ion.

Activity coefficients were calculated from the various forms of the Debye–Huckel equation, including the Debye–Huckel limiting law[13]

$$\log \gamma_i = -Az_i^2 I^{1/2} \tag{3}$$

and the Debye–Huckel expression[1,2]

$$\log \gamma_i = -\frac{Az_i^2 I^{1/2}}{1 + B\mathring{a}I^{1/2}} \tag{4}$$

where γ_i is the activity coefficient of the species of interest, z_i is the charge of the species of interest, I is the ionic strength of the solution, which depends on the charge z_j and concentration C_j of each species present in solution,

$$I = \frac{1}{2} \sum_j C_j z_j^2 \tag{5}$$

A and B are constants dependent on the dielectric constant ε and density d^0 of the solvent and on the temperature T[9]

$$A = 1.825 \times 10^6 (\varepsilon T)^{-3/2} (d^0)^{1/2}$$

$$= 0.509 \text{ in water at } 25°\text{C} \tag{6}$$

$$B = 50.29 (\varepsilon T)^{-1/2} (d^0)^{1/2}$$

$$= 0.33 \text{ in water at } 25°\text{C} \tag{7}$$

and $\mathring{a}$ is the ion-size parameter. Kielland[13,14] assigned a value of 9 to $\mathring{a}$ for the hydrogen ion ($B\mathring{a} = 2.97$); although it may seem a contradiction to associate such a large $\mathring{a}$ value with the smallest ion, the high degree of solvation associated with the proton in aqueous solution provides some justification for this arbitrarily assigned value. The difference between pH calculated according to the Debye–Huckel limiting law and that calculated according to the Debye–Huckel expression increases with increasing ionic strength.

The Debye–Huckel theory predicts a rise in pH with increasing ionic strength, that is, a decrease in the hydrogen ion activity, regardless of which form of the Debye–Huckel equation is employed. The particular form of the equation utilized influences only the magnitude of the predicted pH change.

Table 6.2 contains the observed and calculated pH values for a series of HCl–$K_2Cr_2O_7$ solutions at approximately pH 2.4. These data are also

TABLE 6.2 pH of HCl–$K_2Cr_2O_7$ Solutions at 25°C[a]

[Acid]:[Oxidant][b] Ratio		Observed pH	Standard Deviation	Debye–Huckel Calculated pH: Limiting Law	Debye–Huckel Calculated pH: Expression
∞		2.438	0.007	2.415	2.410
37.58		2.441	0.005	2.416	2.410
18.79		2.441	0.006	2.417	2.411
12.53		2.441	0.008	2.417	2.411
9.396		2.438	0.005	2.418	2.412
7.517		2.440	0.004	2.419	2.413
4.698		2.435	0.006	2.422	2.414
3.758		2.433	0.007	2.423	2.415
1.879		2.432	0.008	2.431	2.420
0.9400		2.430	0.008	2.444	2.427
	ΔpH=	−0.008		+0.029	+0.017

[a] [HCl]=4.149×10^{-3} *M*. pH of standardizing buffers: 2.270, 2.366.
[b] Total oxidant concentration is expressed as dichromate ($0.5C_{Cr}$).

represented graphically in Figure 6.5 as a function of the concentration ratio of hydrogen acid to oxidizing agent. The change in pH over the oxidant concentration range studied, ΔpH, indicates that a decrease in pH is observed with increasing oxidant concentration, contrary to the Debye–Huckel predictions and in agreement with the hypothesis of a unified, all-encompassing acid–base theory that a reducible species enhances solution acidity because it is an acid.

It may be noted at this point that, although the acid : oxidant ratios of Table 6.2 and Figure 6.5 assume the presence of the oxidant as dichromate only, this is not in accord with the equilibria governing the distribution of dichromate and related species in aqueous solution[10,15]

$$HCrO_4^- \rightleftharpoons H^+ + CrO_4^{2-} \tag{8a}$$

$$K_a = \frac{[H^+][CrO_4^{2-}]}{[HCrO_4^-]} = 3.2\times10^{-7} \tag{8b}$$

$$2HCrO_4^- \rightleftharpoons Cr_2O_7^{2-} + H_2O \tag{9a}$$

$$K_d = \frac{[Cr_2O_7^{2-}]}{[HCrO_4^-]^2} = 43 \tag{9b}$$

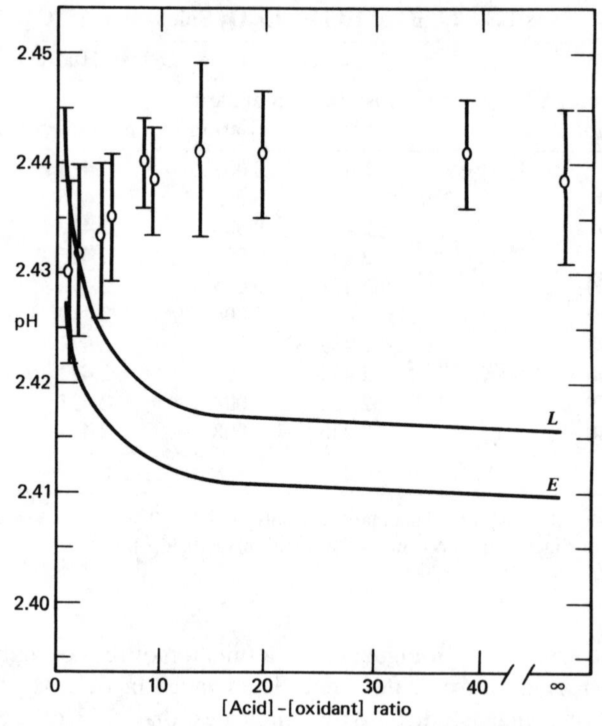

Figure 6.5. pH of HCl–$K_2Cr_2O_7$ solutions at 25°C. [HCl]$=4.149\times10^{-3}$ *M*. Key: $L=$ Debye–Huckel limiting law; $E=$Debye–expression; and ⫯ =experiment (including standard deviation).

Therefore, in any aqueous dichromate solution the analytical chromium concentration C_{Cr} is the sum of the concentrations of the three chromium species in equations (8) and (9):

$$C_{\mathrm{Cr}}=2\left[\mathrm{Cr_2O_7}^{2-}\right]+\left[\mathrm{HCrO_4}^{-}\right]+\left[\mathrm{CrO_4}^{2-}\right] \tag{10}$$

If, as in this case, the solution was prepared by dissolving dichromate, C_{Cr} is equal to twice the original dichromate concentration. Substitution of expressions for bichromate and dichromate concentrations from equations (8) and (9) into equation (10) yields

$$C_{\mathrm{Cr}}=\left(\frac{2K_d[\mathrm{H}^+]^2}{K_a^2}\right)\left[\mathrm{CrO_4}^{2-}\right]^2+\left(\frac{[\mathrm{H}^+]}{K_a}+1\right)\left[\mathrm{CrO_4}^{2-}\right] \tag{11}$$

which may be solved, via the quadratic formula, for $[CrO_4^{2-}]$. Bichromate and dichromate concentrations are then determined by substitution into equations (8) and (9). The results of calculations of the distribution of the dichromate, bichromate, and chromate species for $C_{Cr} = 0.10$ (e.g., 0.05 *M* $K_2Cr_2O_7$) are shown in Table 6.3 and Figure 6.6.

TABLE 6.3 Distribution of Chromium Species in 0.05 *M* $K_2Cr_2O_7$ With Varying pH

pH	$[H^+]$	$[CrO_4^{2-}]$	$[Cr_2O_7^{2-}]$	$[HCrO_4^-]$
0.19999999	0.63095742	0.00000001	0.03561106	0.02877784
0.39999998	0.39810729	0.00000002	0.03561101	0.02877783
0.59999996	0.25118893	0.00000004	0.03561101	0.02877783
0.79999995	0.15848953	0.00000006	0.03561103	0.02877783
0.99999994	0.10000014	0.00000009	0.03561095	0.02877780
1.19999981	0.06309581	0.00000015	0.03561094	0.02877779
1.39999962	0.03981078	0.00000023	0.03561096	0.02877780
1.59999943	0.02511893	0.00000037	0.03561085	0.02877776
1.79999924	0.01584898	0.00000058	0.03561074	0.02877773
1.99999905	0.01000004	0.00000092	0.03561070	0.02877769
2.19999886	0.00630960	0.00000146	0.03561042	0.02877759
2.39999866	0.00398109	0.00000231	0.03561001	0.02877744
2.59999847	0.00251190	0.00000367	0.03560954	0.02877723
2.79999828	0.00158490	0.00000581	0.03560860	0.02877685
2.99999809	0.00100001	0.00000921	0.03560716	0.02877627
3.19999790	0.00063096	0.00001459	0.03560497	0.02877538
3.39999771	0.00039811	0.00002313	0.03560132	0.02877391
3.59999752	0.00025119	0.00003665	0.03559576	0.02877166
3.79999733	0.00015849	0.00005808	0.03558690	0.02876808
3.99999714	0.00010000	0.00009204	0.03557278	0.02876237
4.19999695	0.00006310	0.00014583	0.03555033	0.02875330
4.39999676	0.00003981	0.00023100	0.03551497	0.02873899
4.59999657	0.00002512	0.00036583	0.03545887	0.02871628
4.79999638	0.00001585	0.00057907	0.03537016	0.02868035
4.99999619	0.00001000	0.00091594	0.03523023	0.02862355
5.19999599	0.00000631	0.00144712	0.03500953	0.02853375
5.39999580	0.00000398	0.00228214	0.03466280	0.02839211
5.59999561	0.00000251	0.00358857	0.03412103	0.02816935
5.79999542	0.00000158	0.00561706	0.03328116	0.02782051
5.99999523	0.00000100	0.00872891	0.03199632	0.02727821
6.19999504	0.00000063	0.01341182	0.03007156	0.02644501
6.39999485	0.00000040	0.02024651	0.02728235	0.02518876
6.59999466	0.00000025	0.02974908	0.02344923	0.02335232
6.79999447	0.00000016	0.04199494	0.01860272	0.02079955

TABLE 6.3 (Continued)

pH	$[H^+]$	$[CrO_4^{2-}]$	$[Cr_2O_7^{2-}]$	$[HCrO_4^-]$
6.99999428	0.00000010	0.05607139	0.01320295	0.01752270
7.19999409	0.00000006	0.06988788	0.00816578	0.01378049
7.39999390	0.00000004	0.08114100	0.00438195	0.01009484
7.59999371	0.00000003	0.08884287	0.00209140	0.00697405
7.79999352	0.00000002	0.09352195	0.00092260	0.00463204
7.99999332	0.00000001	0.09621471	0.00038875	0.00300679
8.19999313	0.00000001	0.09774327	0.00015972	0.00192731
8.39999294	0.00000000	0.09861797	0.00006473	0.00122692
8.59999275	0.00000000	0.09906763	0.00002601	0.00077767
8.79999256	0.00000000	0.09922391	0.00001039	0.00049145
8.99999237	0.00000000	0.09935361	0.00000415	0.00031049
9.19999218	0.00000000	0.09982473	0.00000167	0.00019684
9.39999199	0.00000000	0.09671873	0.00000062	0.00012033
9.59999180	0.00000000	0.09897697	0.00000026	0.00007770
9.79999161	0.00000000	0.09040844	0.00000009	0.00004478
9.99999142	0.00000000	0.05677317	0.00000001	0.00001774

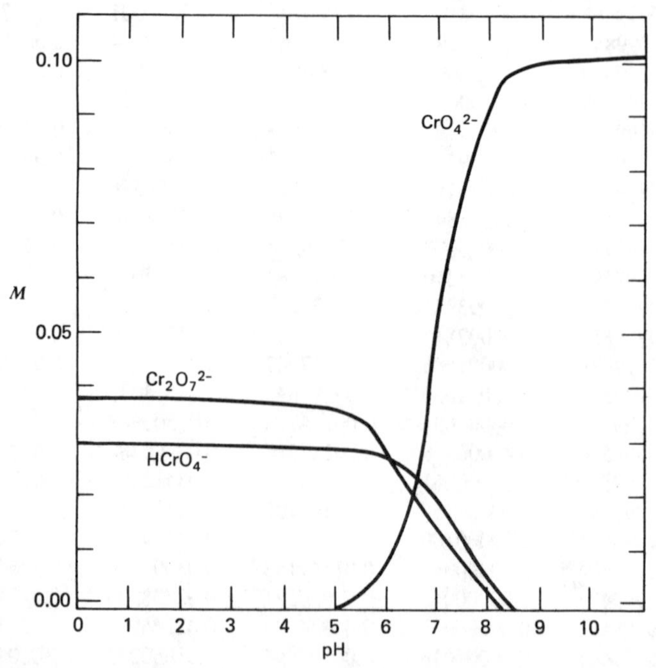

Figure 6.6. Distribution of chromium species in 0.05 M $K_2Cr_2O_7$ with varying pH.

Equation (11) indicates that the distribution is dependent on the two factors pH and C_{Cr}. Table 6.3 indicates that at constant C_{Cr} increasing pH does not significantly alter the distribution up to pH>4; the relative dichromate and bichromate concentrations remain essentially constant, while chromate concentration is negligible at low pH.

The 40-fold variation in C_{Cr} (excluding the dichromate-free solution) over the range of dichromate concentrations studied does affect the distribution of Cr(VI) at a given pH. Calculations using the HCl concentration from Table 6.3 in equation (11) indicate that the bulk of C_{Cr} is present as $HCrO_4^-$ at pH 2.4 and varies from approximately 0.67 C_{Cr} in the most concentrated dichromate solution to 0.98 C_{Cr} in the most dilute one (Figure 6.7).

The question can be raised as to whether it is valid to utilize an acid : oxidant ratio scale in Figure 6.5 based on the presence of the oxidizing agent as dichromate only, if this is not actually the case. Bichromate, $HCrO_4^-$, is almost as good an oxidizing agent as $Cr_2O_7^{2-}$, while CrO_4^{2-} has only weak oxidizing properties (Table 5.1). However, the concentration

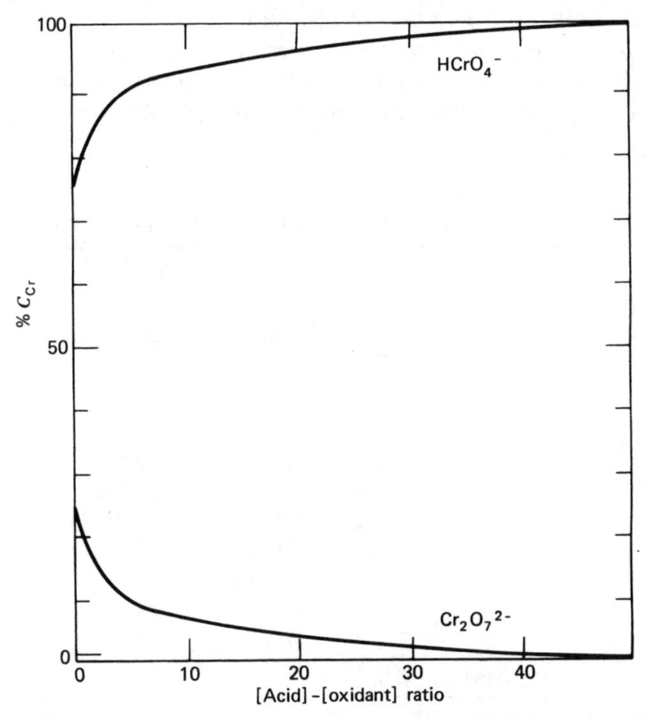

Figure 6.7. Distribution of chromium species at constant pH with varying C_{Cr}. [HCl]= 4.149×10^{-3} *M*.

of the latter is negligible below pH 5, and therefore it is not incorrect to state that the chromium species are quantitatively present as strong oxidants at pH 2.4. While it is simply a matter of calculation to determine the oxidant concentration in terms of the sum of the actual dichromate and bichromate concentrations, important relationships between acidity and specific oxidant structure (to be discussed later) are thus obscured. It is not only convenient, but also advantageous, to utilize a scale in which the total oxidant concentration is expressed in terms of dichromate only; it should be noted that the same arguments apply with equal force to the use of an abcissa in which the total oxidizing agent concentration is expressed in terms of bichromate only (in which case all the numbers on the acid: oxidant ratio scale of Figure 6.5 would be halved).

It may be mentioned at this point that the ionic strengths [equation (5)] used to calculate Debye–Huckel pH values were determined on the basis of $Cr_2O_7^{2-}$ –$HCrO_4^-$ –CrO_4^{2-} distributions calculated via equation (11), and not on the supposition of the presence of only one form of the oxidizing agent. However, the pH calculated in this manner differs only slightly ($\leq$0.005 pH unit) from that calculated on the supposition of the presence of dichromate only, over the range of C_{Cr} studied.

Table 6.4 and Figure 6.8 contain the results of experiments with $K_2Cr_2O_7$ in which $HClO_4$ was substituted for HCl. A pH trend similar to that of the HCl–$K_2Cr_2O_7$ system was observed with increasing oxidizing agent con-

TABLE 6.4 pH of $HClO_4$–$K_2Cr_2O_7$ Solutions at 25°C[a]

[Acid]:[Oxidant][b] Ratio	Observed pH	Standard Deviation	Debye–Huckel Calculated pH: Limiting Law	Debye–Huckel Calculated pH: Expression
∞	2.294	0.006	2.279	2.272
41.04	2.298	0.004	2.280	2.272
20.51	2.298	0.003	2.281	2.273
16.44	2.293	0.005	2.282	2.274
13.68	2.296	0.004	2.282	2.274
10.26	2.296	0.005	2.283	2.274
8.182	2.295	0.005	2.284	2.275
4.101	2.292	0.004	2.288	2.278
2.051	2.287	0.005	2.296	2.283
1.026	2.289	0.005	2.311	2.290
ΔpH	−0.005		+0.032	+0.018

[a] $[HClO_4]=5.750\times10^{-3}$ *M*. pH of standardizing buffers: 2.270, 2.366.
[b] Total oxidant concentration is expressed as dichromate (0.5 C_{Cr}).

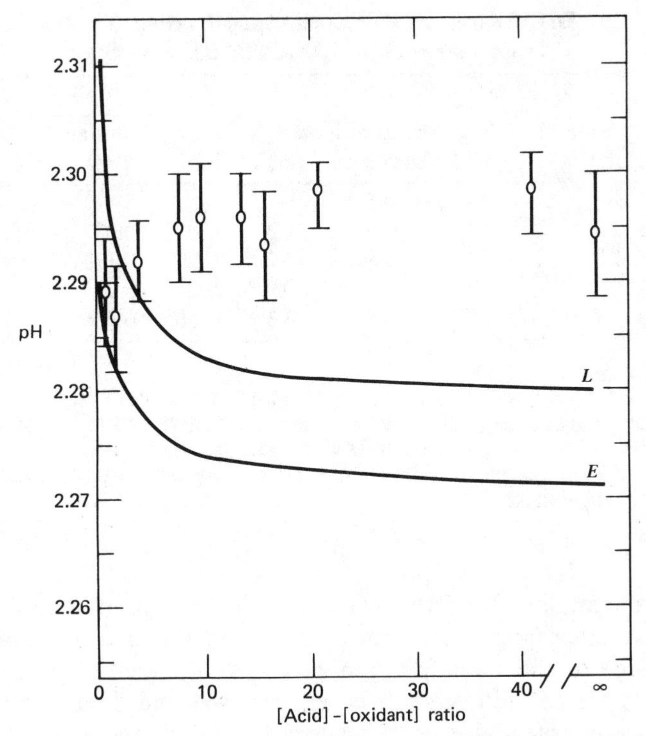

Figure 6.8. pH of $HClO_4$–$K_2Cr_2O_7$ solutions at 25°C. $[HClO_4]=5.750\times10^{-3}$ *M*. Key: L=Debye–Huckel limiting law; E=Debye–Huckel expression; ⫯=experiment (including standard deviation).

centration, indicating that this behavior is not an anomaly characteristic of HCl.

Measurements of the emf E_{cell} of cells containing a series of HCl–$K_2Cr_2O_7$ solutions and an HCl reference solution in the pH 2.3–2.4 range are shown in Table 6.5 and Figure 6.9. These data indicate that E_{cell} increases with increasing oxidant concentration over the range of C_{Cr} studied by more than 3 mV. Since

$$E_{cell} = E_{ref} - E_{ind} \tag{12}$$

where E_{ref} and E_{ind} represent the potentials of the reference and indicator half-cells, respectively, and E_{ref} is constant, E_{ind} must decrease with increasing oxidant concentration. This is due, at least in part, to the decreased activity of hydrogen ion at increased ionic strength (Debye–Huckel effect), but the magnitude of this effect is much greater than the predictions of

TABLE 6.5 Emf Changes and Calculated Liquid-Junction Potential Changes of Cells Containing $HCl–K_2Cr_2O_7$ Solutions at 25°C[a]

[Acid]:[Oxidant][b] Ratio	E_{cell} (mV)	Standard Deviation	ΔE_{cell} (mV)	ΔpH ind[c]	Δe_j (pH units) Limiting Law	Expression
∞	0.4	0.8	—	—	—	—
8.787	0.7	0.9	0.3	0.005	0.082	0.082
3.521	1.6	0.8	1.2	0.020	0.178	0.177
1.753	2.4	0.9	2.0	0.034	0.293	0.290
0.8787	3.7	0.5	3.3	0.056	0.434	0.434

[a]Cell: Glass | $HCl\ (4.701\times10^{-3}\ M),\ K_2Cr_2O_7\ (\times M)$ | $HCl\ (4.701\times10^{-3}\ M)$ | Glass
Indicator half-cell — Reference half-cell

[b]Total oxidant concentration is expressed as dichromate ($0.5C_{Cr}$).

[c]ΔpH_{ind} is the change in the indicator half-cell pH with increasing oxidant concentration $= -\Delta E_{cell}/0.05916$ at 25°C.

either of the forms of the Debye–Huckel equation utilized in Table 6.2 and implies that other factors also contribute to the change in cell potential.

A change in the liquid-junction potential of a cell E_j is sometimes responsible for the difference between observed and predicted pH. The liquid-junction potential may be changed by altering the ionic strength of the sample solution; however, solutions of equal ionic strength containing equal concentrations of an acid but containing different supporting electrolytes are also characterized by unequal E_j values (in cells with identical reference electrodes), which are manifested as differences between the observed pH values of these solutions. The liquid-junction potential depends upon the nature of each individual ionic species in solution, its activity, and its transport number. This is clearly evidenced by the Henderson equation, from which E_j is calculated as the difference between the sums of contributions of all the ions present in the solutions 1 and 2 between which the boundary exists:[9]

$$E_j = \int_1^2 \sum_i \frac{t_i}{z_i}\, dpa_i \qquad (pa_i \text{ units}) \tag{13}$$

where a_i is the activity of species i, z_i is its charge, and t_i is its transport number, which depends on the activities and mobilities u_i of all the ions

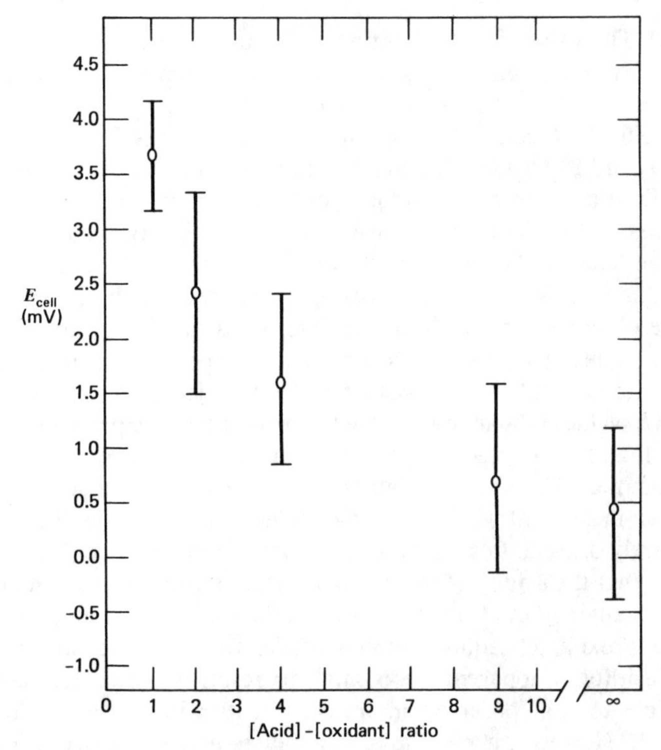

Figure 6.9. Emf of cells containing HCl–$K_2Cr_2O_7$ solutions at 25°C.
Cell: Glass | HCl (4.701×10^{-3} *M*), $K_2Cr_2O_7$ (x *M*)|HCl (4.701×10^{-3} *M*)| Glass
Key: ⫯ = E_{cell} (including standard deviation).

present in solution,[2]

$$t_i = \frac{a_i u_i}{\sum_j a_j u_j} \tag{14}$$

Equivalent ionic conductance at infinite dilution constitutes a reliable measure of relative ionic mobility and may therefore be used to calculate transport numbers.

Calculations of ΔE_j, the change in liquid-junction potential with increasing oxidant concentration, were made for the HCl–$K_2Cr_2O_7$ solutions of Table 6.5 by calculating the contribution to E_j of the indicator half-cell [solution 1 in equation (13)] and subtracting from the result of a similar calculation for the reference half-cell [solution 2 in equation (13), and a constant]. The values for the limiting ionic conductances are from the data tabulated by Robinson and Stokes,[28] except for those of bichromate and dichromate, which were estimated from the extrapolation to infinite dilution of the equivalent conductances, measured at various concentrations, of $Na_2Cr_2O_7$ and $K_2Cr_2O_7$ solutions,[29,30] and then subtracting the appropriate cationic limiting equivalent conductance.* The activity of each ionic species was calculated from both the Debye–Huckel limiting law [equation (3)] and the Debye–Huckel expression with ion-size parameter [equation (4)]. The ion-size parameters of all relevant species required for the latter computation were obtained directly from the data of Kielland,[14] except for dichromate and bichromate, for which $\mathring{a}$ (which is dependent upon the limiting equivalent ionic conductance) was estimated by Kielland's method.

The ΔE_j values tabulated in Table 6.5 predict that the pH should increase by more than four-tenths of a pH unit over the range of oxidant concentrations examined. The observed values of ΔE_{cell} (which is equal to $-\Delta E_{\text{ind}}$), the change in cell emf with increasing oxidant concentration, indicate a rise in pH barely one-eighth as large as ΔE_j over the same range (Figure 6.9). It is evident that the acidity of the oxidizing agent significantly counteracts the effect of changing liquid-junction potential, because ΔE_{cell} and ΔE_j are not even of approximately equivalent magnitude. The magnitude of the effect of oxidant acidity is apparently too small to result in a net decrease in pH comparable to that observed in measurements with a glass–calomel cell (Table 6.2). However, it may be reasonably assumed that the concentrated KCl salt bridge present in the glass–calomel cell is able to suppress, at least partially, the change in liquid-junction potential for a series of dilute, aqueous solutions of similar composition, and that this permits the observation of a decrease in pH. It is for this reason that ΔpH, the change in pH with increasing oxidizing agent concentration, is significant in Table 6.2, although its magnitude is comparable to that of the standard deviation of measurements for each solution.

The decrease in E_{ind} may also be partially attributed to the formation of the chlorochromate ion,[16]

$$HCrO_4^- + H^+ + Cl^- \rightleftharpoons CrO_3Cl^- + H_2O \qquad K = 17 \qquad (15)^\dagger$$

*Since an acidic, aqueous solution of dichromate is actually a dichromate–bichromate mixture, and the mobilities of these ions therefore cannot be determined independently of each other, the result of the extrapolation procedure was considered an "average" limiting equivalent ionic conductance and was applied to the calculation of the transport numbers of both the dichromate and bichromate species.

†Equilibria analogous to equation (15) also occur in aqueous sulfuric and phosphoric acid solutions upon addition of dichromate, but not in perchloric and nitric acids.

which involves the replacement of three ionic species by a single ion, but it is notable that equation (15) predicts a significant rise in pH with increasing oxidant concentration, since hydrogen ion is consumed during the reaction and the magnitude of K is fairly large, which is in contradiction with experimental observation. Thus, despite several reasons (Debye–Huckel effect, change in liquid-junction potential, chlorochromate ion formation) for expecting a large rise in pH with increasing oxidant concentration, an increase in acidity (or, at least, only a small decrease) is observed.

We propose that the decrease in E_{ind} and the simultaneous increase in acidity may be explained by the fact that dichromate and bichromate alter the usual Grotthus mechanism for rapid proton transfer in aqueous solutions (Figure 5.1). These chromium species coordinate with hydrogen (hydronium) ions to form large, bulky species of limited mobility:

$$Cr_2O_7^{2-} + mH_3O^+ \rightleftharpoons (Cr_2O_7 \cdot mH_3O)^{(m-2)+} \quad (16)$$

$$HCrO_4^- + nH_3O^+ \rightleftharpoons (HCrO_4 \cdot nH_3O)^{(n-1)+} \quad (17)$$

This viewpoint is not unreasonable, since an oxygen of the dichromate or bichromate entities should be at least as likely a coordination site for protons as is the oxygen atom of a water molecule, and probably more so, because of the negative charges on these species. The existence of these "hydroniumated" species is at least as possible as the existence of "hydroniumated" Fe(II), which Reynolds and Lumry[18] postulated as a prerequisite to the Grotthus-like electron exchange between ferrous and ferric ions in aqueous acid solutions,

$$Fe^{2+} + H_3O^+ \rightleftharpoons Fe(H_3O)^{3+} \quad (18)$$

and which mandates the close approach and combination of two positively charged species, in direct contradiction with the opinion of Duke,[18] that species of like charge do not interact directly (see Chapter 5 for a complete discussion of these points).

The assumption that each oxygen atom in dichromate or bichromate is capable of coordinating two hydronium ions leads to maximum values of $m=14$ and $n=7$ in equations (16) and (17), respectively. The existence of species with charges as high as those in $(Cr_2O_7 \cdot 14H_3O)^{12+}$ and $(HCrO_4 \cdot 7H_3O)^{6+}$ seems improbable, but it is notable that the experimental curves of Figures 6.5 and 6.8 begin to turn downward, that is, begin to manifest their most visible increases in acidity, at acid:oxidant ratios (where the oxidant is considered to be present totally as dichromate only as a convenience) close to 14:1 (or 7:1 if the oxidant is taken as bichromate). Thus, increased acidity becomes most visible when there is insufficient H_3O^+ present to fully "hydroniumate" all of the available dichromate and bichromate in the sense of the maximum coordination numbers proposed above. This implies that the enhancement of aqueous acidity by these oxidizing agents is related to competition between "hydroniumated" and "non-hydroniumated" chromium species for the available hydronium ion. The oxidizing (electron-accepting) abilities of dichromate and bichromate exert an inductive effect (Figure 6.10) upon coordinated hydronium ions that "loosens" protons from the complex and renders them even more available to rapid proton transfer than in the case of the usual Grotthus mechanism.

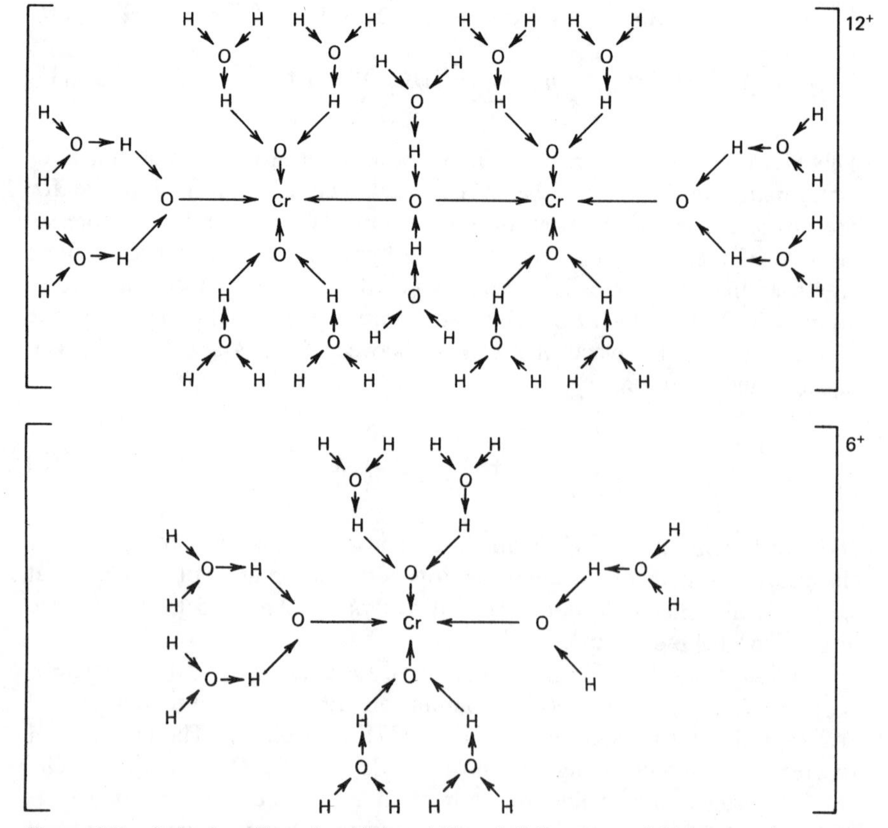

Figure 6.10. Inductive effect of oxidizing ability of dichromate and bichromate ions upon hydronium ion in "hydroniumated" species.

The unusually high degree of freedom afforded protons under these conditions raises their activity and is responsible for the observed increase in acidity with increasing oxidizing acid concentration. The postulated existence of hydroniumated dichromate and bichromate also explains decreased E_{ind} with increasing oxidant concentration, since E_j is dependent on ionic mobility and such large, bulky species are not expected to have high mobilities. It is of interest to note that the hydroniumated species satisfy the Brønsted–Lowry, solvent systems, Lewis, and Usanovich conceptions of an acid, that is, they are proton donating, solvent cation forming, electron pair accepting, and reducible.

Enhancement of acidity by these species is not limited to the very low pH range. Table 6.6 and Figure 6.11 contain the results of a series of experiments with the HCl–$K_2Cr_2O_7$–$LiCl$ system in the pH 4.3–4.4 range. Initial experiments at this pH with solutions containing only HCl and dichromate yielded unstable and irreproducible results. Comparison of the compositions of the sample solutions with those of standard buffers led to the deduction that the low ionic strength of the samples (5×10^{-5}–2×10^{-4}) could be responsible for the difficulties in measurement. Primary standard 0.05 *M* potassium hydrogen phthalate buffer has an ionic strength of approximately 0.05, although its hydrogen ion concentration is only about 10^{-4} *M*. Sample solutions were subsequently prepared with a large excess of supporting electrolyte, which gave stable and reproducible results.

TABLE 6.6 pH of HCl–$K_2Cr_2O_7$–$LiCl$ Solutions at 25°C[a]

[Acid] : [Oxidant][b] Ratio	Observed pH	Standard Deviation	Debye–Huckel Calculated pH: Limiting Law	Debye–Huckel Calculated pH: Expression
∞	4.383	0.011		
39.30	4.374	0.012		
19.65	4.373	0.014		
15.76	4.383	0.011		
13.12	4.368	0.014	4.357	4.307
9.827	4.373	0.009		
7.872	4.370	0.009		
3.954	4.350	0.009		
1.969	4.327	0.008		
0.9847	4.338	0.012		
ΔpH	−0.045			

[a] $[HCl]=5.800\times10^{-5}$ *M*; $[LiCl]=5.551\times10^{-2}$ *M*. pH of standardizing buffer: 4.004.
[b] Total oxidant concentration is expressed as dichromate ($0.5C_{Cr}$).

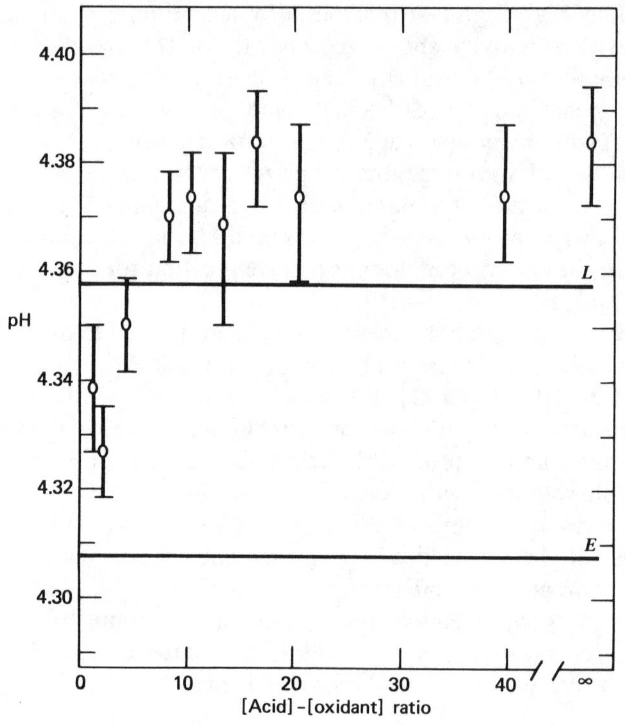

Figure 6.11. pH of $HCl–K_2Cr_2O_7–LiCl$ solutions at 25°C. [HCl]=5.800×10^{-5} M; [LiCl]= 5.551×10^{-3} M. Key: L=Debye-Huckel limiting law; E=Debye-Huckel expression; and ⫯ =experiment (including standard deviation).

The 1000-fold excess of LiCl relative to HCl in each sample solution also provided other advantages. The change in ionic strength over the range of oxidant concentrations examined was small enough (0.1%) to eliminate variations due to the Debye–Huckel effect and changing liquid-junction potential, and permitted the observation of a drop in pH much larger than at pH 2.4. It is interesting to note that, if it is assumed that the Debye–Huckel and liquid-junction potential changes at pH 2.4 obscure part of a significant increase in acidity, the total magnitude of that increase, that is, the sum of the obscured and observed parts, is approximately six-hundredths of a pH unit. Table 6.6 indicates that at pH 4.4, where the interferences have been eliminated by the supporting electrolyte, the increase in acidity is approximately 0.05 pH unit. Taking into account the change in the distribution of chromium species that accompanies a 100-fold decrease in C_{Cr} range [at pH

4.4, chromium (VI) is quantitatively or almost quantitatively present as bichromate over the range of C_{Cr} studied; chromate and dichromate are practically negligible] and the slightly greater oxidizing (and related acidic) properties of dichromate relative to bichromate, it can be seen that the results of Table 6.6 correlate well with those of Table 6.2.

It is not expected that chromate ion, the predominant Cr(VI) species in alkaline solutions, will enhance aqueous acidity, because chromate is not a good oxidizing agent (Table 5.1). The experimental verification of this statement is difficult because the glass electrode lacks selectivity for hydrogen ions in the presence of alkali metal ions at high pH. This results in the "alkaline error" associated with glass electrodes[9,19,20]:

$$\text{pH} = -\log\left(a_H + \sum_{i \neq H} K_{H,i} a_i^{1/z_i}\right) \tag{19}$$

where a_i is the activity of the interfering species i, z_i is the charge of the interfering species, and $K_{H,i}$ is the selectivity constant of the glass electrode for the interfering species over the proton. The alkaline error yields a measured pH lower than the true pH; that is, the change in pH is in the same direction as any enhancement of acidity would be. Consequently, even if enhancement of acidity by chromate were expected, it would be difficult to distinguish from the alkaline error, considering the small magnitude of such enhancement (not more than a few hundredths of a pH unit). Equation (19) indicates that the problem of interference from alkali metal ions can be partially resolved, in general, by substituting divalent or trivalent metal ions for univalent metal ions, since the effective interference varies with a_i^{1/z_i}. However, this is impractical in this case because multivalent cation chromates are only slightly soluble in water. The only freely soluble chromate other than alkali metal chromates is $(NH_4)_2CrO_4$, which is undesirable because it is likely to affect the pH of alkaline media through the action of the NH_4^+ –NH_3 buffer system.

Further evidence supporting the existence of the "hydroniumated" species of equations (15) and (16) was obtained from paper-strip electrophoresis experiments with 0.05 M $K_2Cr_2O_7$ in supporting media of varying pH. The results of these experiments are contained in Table 6.7 and Figure 6.12.

The distance L migrated by an ion under the influence of an applied voltage V in a time t is given by

$$L = atV \tag{20}$$

where a is proportional to the ionic mobility of the migrating species. Holding t and V constant made L directly proportional to a, permitting

TABLE 6.7 Electrophoresis of 0.05 *M* K_2Cr_2O in 0.10 Ionic Strength Supporting Electrolyte Solutions of Varying pH

pH	Average Spot Center Migration Distance (mm)[a]	Average Spot Length (mm)[a]	Tail Length (mm)
1.1	31.1±4.8	4.9±2.6	
1.5	35.3±3.9	15.2±4.7	
1.8	31.3±1.1	14.5±7.8	
	36.6±3.4	14.6±11.9	
2.1	38.8±4.0	4.0±1.0	8.1
3.1	38.1±2.1	4.4±1.3	8.3
4.6	38.8±5.7	4.7±1.5	4.2
	55.3±4.6	9.8±2.8	9.3
4.9	40.0±3.0	6.1±1.3	5.1
	53:3±4.2	6.8±1.7	7.8
7.2	41.6±4.1	9.8±3.9	
	55.8±4.5	10.2±1.8	
8.1	40.0±5.5	9.6±2.3	
	54.6±5.6	11.8±1.9	
10.5	41.5±5.0	8.2±3.9	
	57.0±5.8	11.0±3.7	
11.3	58.8±4.9	12.1±2.0	
12.7	56.4±4.8	6.1±2.7	

[a] Includes ± standard deviation.

direct observation of changes in ionic mobility and, consequently, changes in the nature of the migrating species with changing pH.

Three spots were observed over the range of pH studied. Only one spot (spot *A* in Figure 6.12) is present below pH 1.5; this spot is characterized by a 3-cm migration from its point of origin. A second spot (spot *B*) begins to appear at pH 1.5 as a faint front protruding slightly from spot *A*. The species in spot *B* are therefore slightly more mobile than those in spot *A*. The overlap between the two spots at pH 1.5 is so great that it is impossible to determine the size of spot *B* (or its center and thus its migration distance). At pH 1.8 the separation between spots *A* and *B* is much greater; the two can be distinguished from each other and it can be seen that the characteristic migration distance of spot *B* is 6–7 mm greater than that of spot *A*. Spot *B* is clearly predominant at pH 2.1 but has a tail that is almost 1 cm long, indicating that those species predominant at lower pH have not yet disappeared. This tail does not even begin to disappear until between pH 4 and 5, and vanishes only above pH 5.

A third spot (spot *C*) begins to appear between pH 4 and 5. Spots *B* and *C* are present in approximately equal size from pH 5 through the neutral pH

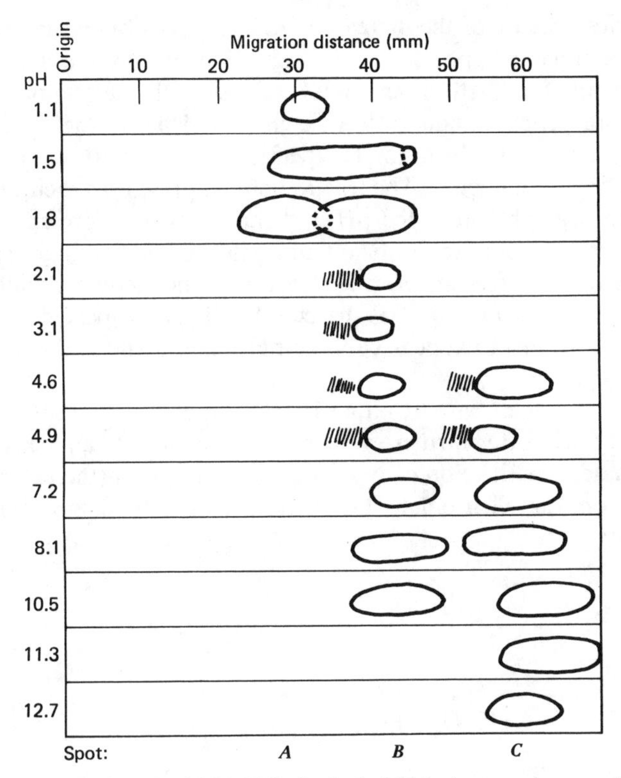

Figure 6.12. Electrophoresis of 0.05 *M* $K_2Cr_2O_7$ in 0.10 ionic strength supporting electrolyte solutions of varying pH.

range. The predominance of spot *C* is beginning to become evident at pH 10.5, and at pH 11.3 only spot *C* remains, although a comparison of spot size at pH 11.3 and 12.7 indicates that some of the species of spot *B* may still be present at pH 11.3.

These three spots cannot be attributed to dichromate, bichromate, and chromate. Equilibrium calculations (which are, however, based on the assumption of the existence of only these three species) indicate that the ratio of dichromate: bichromate remains essentially constant up until pH>4 at constant C_{Cr} (Table 6.3 and Figure 6.6). It is therefore unreasonable to assume that dichromate is responsible for spot *A*, which begins to disappear at about pH 2, and bichromate for spot *B*, which does not begin to appear until pH 1.5. A more logical explanation is that significant equilibria between $Cr_2O_7^{2-}$ and $(Cr_2O_7 \cdot mH_3O)^{(m-2)+}$, and between $HCrO_4^-$ and $(HCrO_4 \cdot nH_3O)^{(n-1)+}$ at low pH, decrease the migration rates of the chromium species because of the increased mass and possibly positive charge of the "hydroniumated" species. Increasing pH decreases the availability of hydronium ion for "hydroniumation," and thus the migration rate rises. We therefore propose that spot *A* is characteristic of the "hydroniumated" dichromate and bichromate species, and spot *B* of their "non-hydroniumated" analogues. The fraction of C_{Cr} present as dichromate does not become negligible until the pH 7–8 range and that present as bichromate does not become insignificant until the pH 8–9 range (Table 6.3), accounting for the presence of spot *B* well into the moderately alkaline pH range. The behavior of spot *C* is in accord with that expected of chromate ion, that is, it begins to appear between pH 4 and 5 and is the only species present at high pH.

The results of acid–base titrations in which a dichromate solution served as the titrant acid offer further evidence of the acidity of aqueous solutions of dichromate. The titration combines the two reactions (the extent of each depending on the dichromate–bichromate–chromate distribution as determined by C_{Cr})

$$Cr_2O_7^{2-} + 2OH^- \rightarrow 2CrO_4^{2-} + H_2O \qquad (21)$$

$$HCrO_4^- + OH^- \rightarrow CrO_4^{2-} + H_2O \qquad (22)$$

and therefore a predicted equivalence point ratio, if the total oxidizing agent concentration is expressed as dichromate, $q_{Cr_2O_7^{2-}}/q_{OH^-} = 0.5000$, or, if the total oxidant concentration is alternatively expressed as bichromate, $q_{HCrO_4^-}/q_{OH^-} = 1.000$, where q_i refers to the quantity (concentration$\times$ volume) of reagent *i*.

The titration of aliquots of 0.2930 *M* NaOH with 0.07696 *M* $K_2Cr_2O_7$ gave an experimental equivalence point ratio $q_{Cr_2O_7^{2-}}/q_{OH^-} = 0.4958 \pm 0.0016$, 0.8% less than the stoichiometric ratio. The equivalence point occurred in the pH 8.9–9.1 region and was relatively sharp and reproducible ($\pm 0.3\%$).

The titration of weighed Na_2CO_3 samples with the same dichromate stock solution was successful only from a purely qualitative standpoint, that is, an inflection point was observed in the pH 8.2–8.4 region (the usual first equivalence point in acid–carbonate titrations). However, this inflection point is generally not very sharp, even if the titrant is a strong acid, and there is a $\pm 3\%$ uncertainty in its exact position. Dichromate is apparently too weak an acid to titrate carbonate to its second equivalence point.

PERMANGANATE

Table 6.8 and Figure 6.13 contain the results of pH measurements of a series of $HClO_4$–$KMnO_4$ solutions at pH 2.3. The behavior of this system agrees well with the predictions of the Debye–Huckel equations. However, these results can also be explained in terms of a unified, all-encompassing acid–base theory if the analogous relationship between MnO_4^- and CrO_4^{2-} is considered. Chromate ion is the least acidic member of the dichromate–bichromate–chromate system, and thus the weakest oxidizing agent of that system. A similar statement can be made for permanganate; the only difference is that the other members of the permanganate system are not as well known as those of the chromate system because structural transitions among the former take place outside the ordinary pH range.

The addition of permanganate to concentrated sulfuric acid results in the formation of the MnO_3^+ ion and/or manganese heptoxide, Mn_2O_7, depending on the concentration of added MnO_4^-.[21] These two species are extremely strong oxidizing agents and decompose quickly (within a day of preparation) and sometimes violently in air to MnO_2 and O_2; exposure to moist air often results in explosion.[21-23] The MnO_3^+ ion and Mn_2O_7 are stored with relative stability and safety in CCl_4, but even in these solutions the oxidizing power and acidity of these species are manifest. Exposure of a solution of MnO_3^+ in CCl_4 to moisture results in slow hydrolysis to yield permanganate. Manganese heptoxide in the same solvent decomposes slowly to MnO_3Cl, MnO_2, and $COCl_2$.[23]

TABLE 6.8 pH of $HClO_4$–$KMnO_4$ Solutions at 25°C[a]

[Acid]:[Oxidant] Ratio	Observed pH	Standard Deviation	Debye–Huckel Calculated pH: Limiting Law	Debye–Huckel Calculated pH: Expression
∞	2.298	0.004	2.283	2.276
40.75	2.299	0.002	2.283	2.276
20.37	2.304	0.002	2.284	2.276
16.36	2.303	0.002	2.284	2.277
13.58	2.307	0.003	2.284	2.277
10.18	2.306	0.002	2.285	2.277
8.141	2.307	0.001	2.285	2.277
4.081	2.303	0.002	2.287	2.279
2.040	2.309	0.002	2.291	2.281
1.021	2.312	0.003	2.298	2.285
ΔpH=	+0.014		+0.015	+0.009

[a] $[HClO_4]=5.697\times10^{-3}$ *M*. pH of standardizing buffers: 2.270, 2.366.

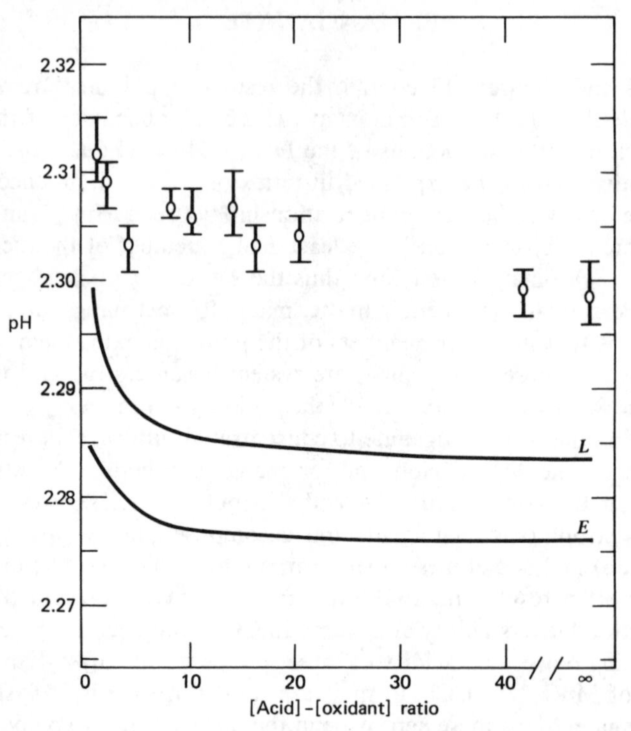

Figure 6.13. pH of $HClO_4$–$KMnO_4$ solutions at 25°C. $[HClO_4]=5.697\times10^{-3}$ *M*. Key: L = Debye–Huckel limiting law; E = Debye–Huckel expression; ⫯ = experiment (including standard deviation).

The MnO_3^+ ion and Mn_2O_7 may be regarded as analogues of CrO_3 and $Cr_2O_7^{2-}$, respectively, but the former are much stronger acids and oxidants than the latter. Upon conversion of MnO_3^+ and Mn_2O_7 to MnO_4^-, both oxidizing power and conventional acidity decrease, just as in the conversion of dichromate to chromate. However, since the oxidizing power of each Mn(VII) species surpasses significantly that of its Cr(VI) analogue, and the conventional acidities of MnO_3^+ and Mn_2O_7 greatly surpass those of their chromium analogues, it seems logical to consider permanganate to be much more acidic than chromate. This is supported by Usanovich's assertion of the equivalence of permanganate acidity and oxidizing ability,[24] which may also be interpreted in terms of Lewis acid–base theory[25] (see Chapter 5). Thus, although permanganate appears to be too weak an acid to counter the Debye–Huckel effect in aqueous solution, this does not mean that permanganate possesses no acidic tendencies whatsoever. Even chromate, a much weaker acid/oxidant, can, in principle, manifest acidity in the presence of a sufficiently strong base/reducing agent, for example, metallic sodium.

It must be noted at this point that, in accord with the electronic acid–base theory, no single order of acid or base strength exists, and that relative acidity is dependent on the identity of the reference base. Permanganate is a stronger acid than dichromate towards the electron as a reference base, as the difference in oxidizing power indicates, but weaker towards hydroxide as a reference base. Consequently, potentiometric titrations of NaOH with $KMnO_4$ were not expected to yield results analogous to those of $K_2Cr_2O_7$–NaOH titrations; attempts to perform such titrations verified this prediction.

The tendency of permanganate to decompose in aqueous solution, which may be regarded as one consequence of its acidity, interfered with some of the attempted experiments with this oxidizing agent. Some of the sample solutions used in the pH experiment at pH 2.3 evidenced traces of MnO_2 after several days, in spite of precautions taken (deaeration, storage in dark under a parafilm seal), but there was no discernible effect on the pH of these solutions. An attempted extension of the pH experiments to pH 4.3 was impossible owing to total decomposition of permanganate within hours of preparation. Attempts to observe the presence, if any, of hydroniumated species via paper-strip electrophoresis at varying pH were ruined by decomposition literally as one watched in both acidic and basic supporting media on both types of paper strips employed.

FERRIC ION

Ferric ion is not as powerful an oxidizing agent as dichromate or permanganate, but is nevertheless of particular interest because of its well-established effect on the acidity of aqueous solutions. It is well known that ferric hydroxide (or, more correctly, hydrated ferric oxide) precipitates in aqueous ferric salt solutions at pH>3, with a resultant increase in the acidity of the solution. Even at lower pH the presence of Fe(III) increases solution acidity via the following reactions:[21]

$$[Fe(H_2O)_6]^{3+} \rightleftharpoons [Fe(H_2O)_5(OH)]^{2+} + H^+ \qquad K_{a1} = 8.9 \times 10^{-4} \tag{23}$$

$$[Fe(H_2O)_5(OH)]^{2+} \rightleftharpoons [Fe(H_2O)_4(OH)_2]^{+} + H^+ \qquad K_{a2} = 5.5 \times 10^{-4} \tag{24}$$

$$2[Fe(H_2O)_6]^{3+} \rightleftharpoons [Fe(H_2O)_4(OH)_2Fe(H_2O)_4]^{4+} + 2H^+ \qquad K_{a3} = 1.2 \times 10^{-3} \tag{25}$$

The oxidation of ferrous complexes containing other protonic ligands to their ferric counterparts has also been shown to increase the acidity of the complex, as manifested by the increasing ability to release protons.[27]

These considerations, along with the fact that an especially large Debye–Huckel effect is expected for solutions containing Fe(III) because of the effect of trivalent ions upon ionic strength [equation (4)] and consequently upon hydrogen ion activity, make the study of Fe(III) systems significant in terms of the proposed unified acid–base concept.

The results of a study of the HCl–$FeCl_3$ system at pH 1.3 (a level of aqueous acidity sufficiently high to prevent precipitation of the hydrous oxide) are given in Table 6.9 and in Figure 6.14. A decrease in pH of a magnitude comparable to that of the HCl–$K_2Cr_2O_7$ and $HClO_4$–$K_2Cr_2O_7$ systems at pH 2.3–2.4 (Tables 6.2 and 6.4, respectively) is observed, in contradiction with Debye–Huckel predictions. It must be noted at this point that the full magnitude of the effect of increasing ferric ion concentration on acidity is not reflected by the results of Table 6.9, because Fe(III) shows a strong tendency to form complexes with chloride ion via the following series of reactions:[17,18,21]

$$Fe^{3+} + Cl^- \rightleftharpoons FeCl^{2+} \qquad K_1 = 30 \tag{26}$$

$$FeCl^{2+} + Cl^- \rightleftharpoons FeCl_2^{+} \qquad K_2 = 4.5 \tag{27}$$

$$FeCl_2^{+} + Cl^- \rightleftharpoons FeCl_3 \qquad K_3 = 0.1 \tag{28}$$

$$FeCl_3 + Cl^- \rightleftharpoons FeCl_4^{-} \qquad K_4 = 0.01 \tag{29}$$

TABLE 6.9 pH of HCl–$FeCl_3$ Solutions at 26°C[a]

[Acid]:[Oxidant][b] Ratio	Observed pH	Standard Deviation	Debye–Huckel Calculated pH: Limiting Law	Debye–Huckel Calculated pH: Expression
∞	1.332	0.007	1.384	1.336
37.85	1.339	0.006	1.390	1.338
18.95	1.340	0.010	1.395	1.339
15.14	1.337	0.006	1.397	1.340
12.56	1.341	0.006	1.400	1.341
9.473	1.341	0.009	1.404	1.342
7.571	1.338	0.002	1.408	1.344
3.777	1.339	0.010	1.425	1.348
1.889	1.322	0.008	1.450	1.355
0.9492	1.322	0.006	1.466	1.358
ΔpH=	−0.010		+0.082	+0.022

[a] $[HCl] = 5.424 \times 10^{-3}$ *M*. pH of standardizing buffer: 1.679

[b] Total oxidant concentration is expressed by C_{Fe}.

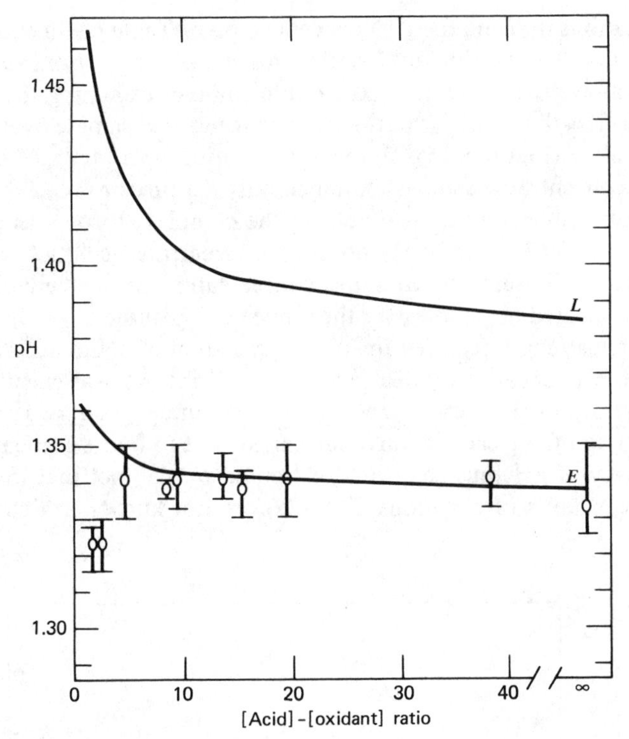

Figure 6.14. pH of HCl–$FeCl_3$ solutions at 25°C. [HCl]$=5.424\times10^{-2}$ *M* Key: *L*=Debye–Huckel limiting law; *E*=Debye–Huckel expression; $\updownarrow$=experimental (including standard deviation).

Thus there is significantly less "free" (in the sense of not being complexed with chloride) ferric ion present than the analytical concentration of iron, C_{Fe}, dissolved in each sample:

$$C_{Fe}=[Fe^{3+}]+[FeCl^{2+}]+[FeCl_2{}^{+}]+[FeCl_3]+[FeCl_4{}^{-}]$$

$$=[Fe^{3+}]\big(1+K_1[Cl^-]+K_2K_1[Cl^-]^2$$

$$+K_3K_2K_1[Cl^-]^3+K_4K_3K_2K_1[Cl^-]^4\big) \tag{30}$$

The amount of chloride ion remaining uncomplexed is therefore

$$[Cl^-]=C_{Cl}-[FeCl^{2+}]-2[FeCl_2{}^{+}]-3[FeCl_3]-4[FeCl_4{}^{-}] \tag{31}$$

Calculations utilizing the HCl concentration of Table 6.9 in equation (28) indicate that the fraction of Fe(III) present as "free" iron varied from approximately $0.31C_{Fe}$ in the most dilute sample (excluding the zero iron sample) to less than $0.07C_{Fe}$ in the most concentrated sample over the range of C_{Fe} studied (Figure 6.15), although C_{Fe} varied by a factor of 40, that is, the most concentrated sample contained only 8.5 times as much "free" iron as the most dilute sample. The bulk of the complexed iron was present as $FeCl^{2+}$ and $FeCl_2{}^+$; only in the most concentrated solution was $FeCl_3$ present in >1% quantity, and the concentration of the tetrachloro-iron complex remained negligible over the range of C_{Fe} studied.

The ionic strength required for the computation of pH in accord with the Debye–Huckel equations for each solution in Table 6.9 was calculated from concentrations of the various iron species determined from equations (26)–(30). Although these calculations were carried out to four significant figures, the accuracy of pH thus calculated is limited by the fact that the complex stability constants of equations (26)–(29) are not known to this degree of

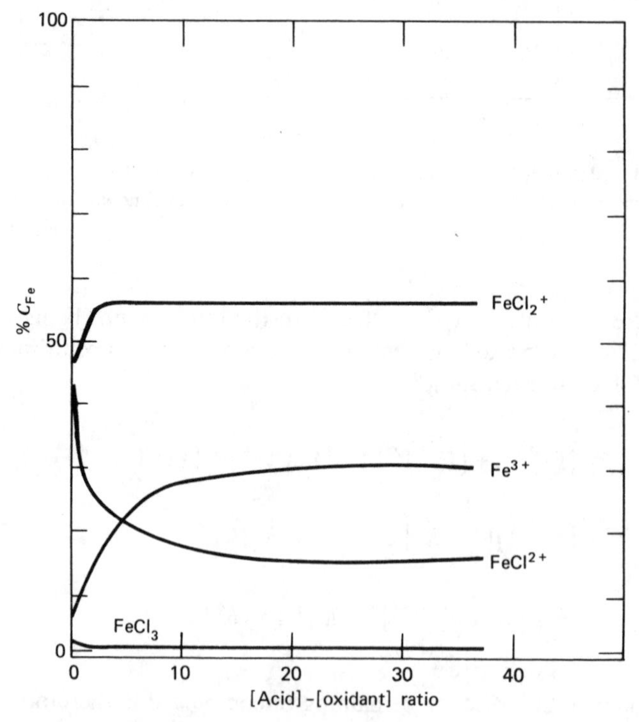

Figure 6.15. Distribution of chloro-iron species at constant pH with varying C_{Fe}. [HCl]= 5.424×10^{-2} *M*.

accuracy. However, this consideration does not affect the direction of the Debye–Huckel predictions and probably affects their magnitude only slightly. The dissociation equilibria of equations (23)–(25) do not proceed to an appreciable enough extent to influence the ionic strength at pH 1.3, although these equilibria are responsible for the characteristic yellow-orange color of Fe(III) solutions and for the observed increase in acidity.

A noticeable drop in pH with time was observed in the most concentrated iron solution during the approximately 8-week period of this study. The pH of this solution remained fairly constant for the first three weeks (and the mean of these results is given in Table 6.9), then decreased to pH 1.30 for the next three weeks, and dropped to pH 1.27 during the final two weeks. No comparable pH changes were observed for the other sample solutions. It is conceivable that this effect is due to the slow formation of even larger aggregates than that of equation (23) above a certain level of "free" iron concentration, the existence of which is compatible with the well-known uncertainty in the composition of hydrated ferric oxide precipitates.

Although Fe(III) complexes to some extent with all the anions of strong acids, it combines with perchlorate to the least extent:[18]

$$Fe^{3+} + ClO_4^- \rightleftharpoons FeClO_4^{2+} \qquad K = 2.10 \tag{32}$$

TABLE 6.10 pH of $HClO_4$–$Fe(ClO_4)_3$ Solutions at 25°C[a]

[Acid]:[Oxidant][b] Ratio	Observed pH	Standard Deviation	Debye–Huckel Calculated pH: Limiting Law	Debye–Huckel Calculated pH: Expression
∞	1.388	0.007	1.457	1.415
42.70	1.385	0.006	1.464	1.418
21.37	1.388	0.009	1.470	1.420
17.16	1.391	0.006	1.474	1.421
14.24	1.389	0.008	1.477	1.422
10.69	1.390	0.007	1.482	1.424
8.540	1.388	0.006	1.488	1.426
4.282	1.368	0.006	1.513	1.433
2.141	1.357	0.008	1.552	1.442
1.071	[c]		1.610	1.453
ΔpH=	−0.031		+0.153	+0.038

[a] $[HClO_4] = 4.475 \times 10^{-2}$ *M*. pH of standardizing buffer: 1.679.

[b] Total oxidant concentration is expressed by C_{Fe}.

[c] pH value not reproducible.

The results of a study of the $HClO_4$–$Fe(ClO_4)_3$ system are shown in Table 6.10 and Figure 6.16. The higher concentration of "free" iron (anywhere from 3 to 11 times as high as in equimolar $FeCl_3$ solutions in the range of C_{Fe} studied; see Figure 6.17) in the perchlorate system permits increased participation by equilibria (23)–(25); consequently a greater increase in acidity is expected than in the ferric chloride system, although a greater Debye–Huckel effect is also expected.

The observed decrease in pH is three times as large as that of the HCl–$FeCl_3$ system. This pH change does not take into account the acidity of the most concentrated ferric perchlorate system, which, like its chloride counterpart, experienced a change in pH with time. However, the much greater concentration of "free" iron in the perchlorate solution accelerated those processes that increase solution acidity to the extent that no reproducible measurements could be obtained; that is, there was a continuous drop in pH from measurement to measurement, and sometimes even during

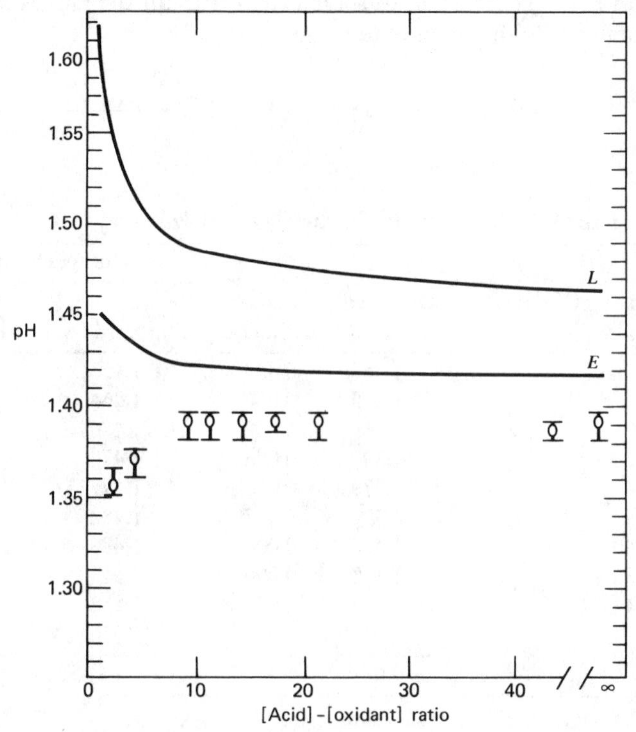

Figure 6.16. pH of $HClO_4$-$Fe(ClO_4)_3$ solutions at 25°C. $[HClO_4]=4.475\times10^{-2}$ *M*. Key. *L*=Debye–Huckel limiting law; *E*=Debye–Huckel expression; Ī=experimental (including standard deviation).

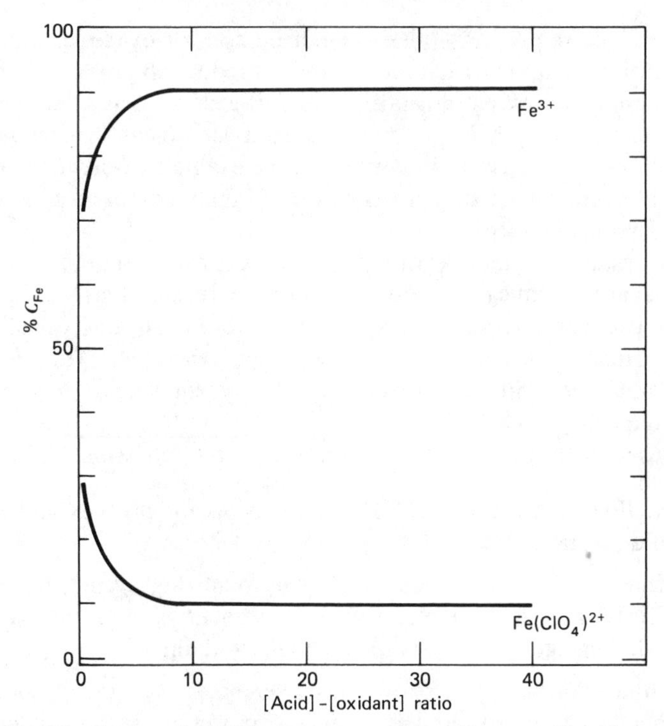

Figure 6.17. Distribution of perchloratoiron species at constant pH with varying C_{Fe}. $[HClO_4]=4.475\times10^{-2}$ *M*.

a single measurement. Therefore no experimental result is given for this solution in Table 6.10.

Paper-strip electrophoresis experiments with ferric species in supporting media of varying pH were limited to the pH 1.1–2.4 range because evidence of precipitation was observed at pH$\geqslant$2.4. Only one spot was observed, with little evidence of tailing, indicating that all of the Fe(III) species present have approximately equal ionic mobilities or are in rapid equilibrium with each other. The positive charge associated with these species makes "hydroniumation" less likely in this case than in the case of dichromate.

CONCLUSIONS

The inclusion of oxidation–reduction processes in acid–base chemistry, first proposed by Usanovich, does not merely extend the scope of the latter but offers a fresh perspective on the general nature of chemical processes. According to this viewpoint, one measure of relative acid strength is the extent of reaction of the acid with electrons (oxidizing power). The electron plays the same role as any other reference base, for example, water in the determination of aqueous acidities. Standard reduction potentials are consequently simply a ranking of acidity against the electron as a reference base. This order of strength is not universal and is subject to change with a change in reference base, as illustrated by the example of permanganate and dichromate. Analogous statements can be formulated concerning the basic nature of reducing agents.

As a result of the integration of redox phenomena into acid–base reactions, any chemical process can now be regarded as an interaction between two species, an acid and a base (although traditionalists may propose other names to distinguish between reactants). Thus a unified, all-encompassing acid–base theory is actually equivalent to a theory of general reactivity.

A unified acid–base theory is supported by the following:

1. Analogous Grotthus transfer mechanisms for protons and electrons in aqueous solution.
2. Enhancement of aqueous acidity by oxidizing agents, for example, Fe(III) and dichromate, and enhancement of aqueous basicity by reducing agents, for example, metallic sodium.
3. Enhancement of the oxidizing power of a given redox couple with increasing acidity, and of reducing power of a given couple with increasing basicity (Table 5.1).

It is believed that oxidizing agents are able to manifest acidity in aqueous solution by altering the Grotthus mechanism for proton transfer and coordinating with hydronium ions. The electron-accepting tendency of the oxidizing agent exerts an inductive effect on the electrons of these hydronium ions, permitting the environment of some of the protons of hydroniumated oxidants to approach that of the free proton and thereby increasing hydrogen ion activity. Even those oxidizing agents that are too weak to manifest this tendency, for example, permanganate, can be shown to have some acidic properties.

REFERENCES

1. Blaedel, W. J., and Meloche, V. W., "Elementary Quantitative Analysis, Theory and Practice," Row, Peterson, and Co., White Plains, New York, 1957.
2. Kolthoff, I. M., and Sandell, E. B., "Textbook of Quantitative Inorganic Analysis," 3rd ed., Macmillan, New York, 1952.
3. Day, Jr., R. A., and Underwood, A. L., "Quantitative Analysis," 3rd ed., Prentice-Hall, Englewood Cliffs, New Jersey, 1974.

4. Waser, J., "Quantitative Chemistry," rev. ed., Benjamin, New York, 1964.
5. Bates, R. G., Hamer, W. J., Manov, G. G., and Acree, S. F., *J. Nat. Bur. Stand.*, **29**, 183 (1942) RP 1495.
6. Scott, W. W., "Standard Methods of Chemical Analysis," Vol. I, 4th ed., Van Nostrand, New York, 1925.
7. Treadwell, F. P., and Hall, W. T., "Analytical Chemistry," Vol. II, Wiley, New York, 1942.
8. Fritz, J. S., and Schenk, Jr., G. H., "Quantitative Analytical Chemistry," 3rd ed., Allyn and Bacon, Boston, 1974.
9. Bates, R. G., "Determination of pH, Theory and Practice," 2nd ed., Wiley, New York, 1973.
10. Weast, R. C. (Ed.), "Handbook of Chemistry and Physics," 51st ed., The Chemical Rubber Co., Cleveland, Ohio, 1970.
11. Durst, R. A., "Standard Reference Materials: Standardization of pH Measurements," National Bureau of Standards Special Publication 260-53, United States Department of Commerce/National Bureau of Standards, Washington, D.C., 1975.
12. Willard, H. H., Merritt, Jr., L. L., and Dean, J. A., "Instrumental Methods of Analysis," 5th ed., Van Nostrand, New York, 1974.
13. Lee, T. S., "Treatise on Analytical Chemistry," Part I, Vol. I, Kolthoff, I. M., Elving, P. J., and Sandell, E. B., Eds., Wiley-Interscience, New York, 1959, Chap. 7.
14. Kielland, J., *J. Am. Chem. Soc.*, **59**, 1675 (1937).
15. Neuss, J. D., and Rieman, W., *J. Am. Chem. Soc.*, **56**, 2238 (1934).
16. Haight, Jr., G. P., Richardson, D. C., and Coburn, N. H., *Inorg. Chem.*, **3**, 1777 (1964).
17. Reynolds, W. L.,.and Lumry, R. W., *J. Chem. Phys.*, **23**, 2460 (1955).
18. Duke, F. R., *J. Am. Chem. Soc.*, **70**, 3975 (1948).
19. Koryta, J., *Anal. Chim. Acta*, **61**, 329 (1972).
20. Morf, W. E., Ammann, D., Pretsch, E., and Simon, W., *Pure Appl. Chem.*, **36**, 421 (1973).
21. Cotton, F. A., and Wilkinson, G., "Advanced Inorganic Chemistry," 3rd ed., Wiley, New York, 1973.
22. Englebrecht, A., and Grosse, A. V., *J. Am. Chem. Soc.*, **76**, 2042 (1954).
23. Briggs, T. S., *J. Inorg. Nucl. Chem.*, **30**, 2866 (1968).
24. Usanovich, M. I., *J. Gen. Chem. USSR*, **21**, 2181 (1951).
25. Wiberg, K., and Stewart, R., *J. Am. Chem. Soc.*, **78**, 1214 (1956).
26. Mohanty, J. G., and Chakravorty, A., *Inorg. Chem.*, **16**, 1561 (1977).
27. Yatsimirskii, K. B., and Vasil'ev, V. P., "Instability Constants of Complex Compounds," Consultant Bureau Enterprises, New York, 1960.
28. Robinson, R. A., and Stokes, R. H., "Electrolyte Solutions," 3rd ed., Butterworths Scientific Publications, London, 1959, p. 463.
29. Washburn, E. W., Ed., "International Critical Tables of Numerical Data, Physics, Chemistry, and Technology," vol. VI, National Research Council, McGraw-Hill, New York, 1929, pp. 251, 253.
30. Jones, H. C., et al., "The Electrical Conductivity, Dissociation, and Temperature Coefficients of Conductivity from Zero to Sixty-Five Degrees of Aqueous Solutions of Salts and a Number of Organic Acids," Publication #1700, The Carnegie Institute of Washington Publications, Washington, D.C., p. 1912.

AUTHOR INDEX

SUBJECT INDEX